For my loving wife, Evelyn.

Louis H. Turcotte

To Emma, who is the light of my life.

Howard B. Wilson

Contents

1 Axially Loaded Systems — 1
  1.1 Analysis of Two Rods Subjected to Gravity Loading — 1
    1.1.1 Problem Solution — 1
    1.1.2 Program to Analyze Two Rods Subjected to Gravity Loading — 3
    1.1.3 Exercises — 5
    1.1.4 Program Output and Code — 9
  1.2 Analysis of a Two Member Truss — 12
    1.2.1 Problem Solution — 12
    1.2.2 Program to Analyze a Two Member Truss — 15
    1.2.3 Exercises — 17
    1.2.4 Program Output and Code — 18

2 Stress and Strain — 24
  2.1 Plate Analysis Using Hooke's Law — 24
    2.1.1 Program to Calculate Stresses or Strains in a Plate — 25
    2.1.2 Exercises — 26
    2.1.3 Program Output and Code — 26
  2.2 Stresses in the Bolts of a Flange Coupling — 31
    2.2.1 Program to Determine Bolt Stresses in a Flange Coupling — 32
    2.2.2 Exercises — 35
    2.2.3 Program Output and Code — 35

3 Statically Indeterminate Systems — 39
  3.1 Multiple Rods with a Common Load Point — 39
    3.1.1 Program to Analyze Multiple Rods With Common Load Point — 41
    3.1.2 Exercises — 42
    3.1.3 Program Output and Code — 42
  3.2 Rigid Bar Suspended by Cables — 45
    3.2.1 Program to Analyze a Rigid Bar Suspended by Cables — 47
    3.2.2 Exercises — 50
    3.2.3 Program Output and Code — 50

3.3 Axially Loaded Shaft   57
    3.3.1 Program to Analyze an Axially Loaded Shaft   59
    3.3.2 Exercises   59
    3.3.3 Program Output and Code   60

4 Geometrical Properties of Polygons   65
4.1 Properties of General Areas   65
    4.1.1 Parallel-Axis Theorem   66
    4.1.2 Rotation of Inertia   68
4.2 Geometrical Properties of Polygons   70
    4.2.1 Program to Determine Geometrical Properties   74
    4.2.2 Using a Polygon to Approximate a Circle   78
    4.2.3 Exercises   78
    4.2.4 Program Output and Code   80

5 Transformation of Stress and Strain   92
5.1 Transformation of Plane Stress   92
    5.1.1 Rotation of Plane Stress   93
    5.1.2 Program for Plane Stress   95
    5.1.3 Exercises   96
    5.1.4 Program Output and Code   101
5.2 Transformation of Plane Strain   119
    5.2.1 Rotation of Strain   119
    5.2.2 Exercises   121
5.3 Strain Rosettes   121
    5.3.1 Basic Mathematical Relationships   122
    5.3.2 The Rectangular Rosette   123
    5.3.3 The Delta Rosette   124
    5.3.4 The T-delta Rosette   126
    5.3.5 Compact Form of Rosette Relationships for Principal Stress   126
    5.3.6 Program to Determine Rosette Strains and Stresses   127
    5.3.7 Exercises   127
    5.3.8 Program Output and Code   128

6 Stresses in Members of Polygonal Cross Section   133
6.1 Moment Angle Required to Induce Maximum Normal Stress   133
    6.1.1 Program to Determine Angle Producing Maximum Normal Stress   135
    6.1.2 Exercises   136
    6.1.3 Program Output and Code   138
6.2 Kern of a Compression Member   143
    6.2.1 Mathematical Relationships   144
    6.2.2 Kern Construction   146
    6.2.3 Program to Determine the Kern   148

Computer Applications in Mechanics of Materials Using MATLAB®

LOUIS H. TURCOTTE
Mississippi State University

HOWARD B. WILSON
University of Alabama

Prentice Hall
New Jersey 07458

Library of Congress Cataloging-in-Publication Data

Turcotte. Louis H.
    Computer Applications in Mechanics of Materials Using MATLAB /
  Louis H. Turcotte,, Howard B. Wilson
      p.   cm.
  Includes bibliographical references and index.
  ISBN 0-13-749060-7
    1. Materials—Mathematical models. 2. MATLAB. 3. Numerical
analysis—Data processing. 4. Computer-aided engineering.
I. Wilson, H. B. (Howard B.)   II. Title.
TA405.T787   1998
620.1'1'015118—dc21   97–17370
                                                                                                        CIP

Acquisitions editor: *BILL STENQUIST*
Editor-in-chief: *MARCIA HORTON*
Managing editor: *BAYANI MENDOZA DE LEON*
Director of production and manufacturing: *DAVID W. RICCARDI*
Production editor: *KATHARITA LAMOZA*
Cover design: *BRUCE KENSELAAR*
Manufacturing buyer: *JULIA MEEHAN*
Editorial assistant: *MEG WEIST*

©1998 by Prentice-Hall, Inc.
Simon & Schuster / A Viacom Company
Upper Saddle River, New Jersey 07458

All rights reserved. No part of this book may be
reproduced, in any form or by any means,
without permission in writing from the publisher.

The author and publisher of this book have used their best efforts in preparing this book. These efforts include the development, research, and testing of the theories and programs to determine their effectiveness. The author and publisher make no warranty of any kind, expressed or implied, with regard to these programs or the documentation contained in this book. The author and publisher shall not be liable in any event for incidental or consequential damages in connection with, or arising out of, the furnishing, performance, or use of these programs.

This book was prepared using L^AT_EX and T_EX. It was published from T_EX files prepared by the authors, and printed from camera-ready copy.

T_EX is a trademark of the American Mathematical Society. MATLAB and SIMULINK are tademarks of The MathWorks, Inc.

Printed in the United States of America

10 9 8 7 6 5 4 3 2 1

ISBN   0-13-749060-7

Prentice-Hall International (UK) Limited, London
Prentice-Hall of Australia Pty. Limited, Sydney
Prentice-Hall Canada Inc., Toronto
Prentice-Hall Hispanoamericana, S.A., Mexico
Prentice-Hall of India Private Limited, New Delhi
Prentice-Hall of Japan, Inc., Tokyo
Simon & Schuster Asia Pte. Ltd., Singapore
Editora Prentice-Hall do Brasil, Ltda., Rio de Janeiro

	6.2.4	Exercises	150
	6.2.5	Program Output and Code	152
6.3	Distribution of Shear Stresses Due to Bending		156
	6.3.1	Program to Determine the Shear Flow in a Cross Section	157
	6.3.2	Program to Plot the Distribution of Shear Stresses	157
	6.3.3	Exercises	162
	6.3.4	Program Output and Code	162
6.4	Bending of a Curved Beam with a Polygonal Cross Section		173
	6.4.1	Bending Stress in a Curved Beam	173
	6.4.2	Algorithm for Geometry Integral	175
	6.4.3	Traditional Forms for Curved Beam Analysis	176
	6.4.4	Program to Calculate Normal Stress in a Curved Beam	177
	6.4.5	Exercises	177
	6.4.6	Program Output and Code	178
6.5	Shear Center for Open Thin-Walled Members		184
	6.5.1	Geometrical Properties	184
	6.5.2	Shear Center	186
	6.5.3	Program to Calculate the Shear Center	188
	6.5.4	Exercises	188
	6.5.5	Program Output and Code	189

7 Flexural Analysis and Deflection of Beams Using Discontinuity Functions 199

7.1	Review of Discontinuity Functions		199
7.2	Application of Discontinuity Functions to Beams		201
7.3	Approximate Beam Analysis Using Superposition		201
	7.3.1	Equations for a Concentrated Load	202
	7.3.2	Example Using A Uniformly Distributed Load	204
	7.3.3	Example with Multiple Loadings	209
	7.3.4	Exercises	213
	7.3.5	Program Output and Code	216
7.4	Analysis of Single Span Beams Using Discontinuity Functions		226
	7.4.1	Program Development	226
	7.4.2	Example with Multiple Loadings	233
	7.4.3	Exercises	235
	7.4.4	Program Output and Code	235

8 Additional Topics 246

8.1	Stresses in an Eccentric Shear Connection		246
	8.1.1	Development of Stress Relationships	246
	8.1.2	Program to Calculate the Bolt Stresses	250
	8.1.3	Program Output and Code	250
8.2	Combined Axial and Flexural Loading of a Beam Column		254
	8.2.1	Program to Analyze an Axially Loaded Beam	256

		8.2.2	Exercises	256
		8.2.3	Program Output and Code	258
	8.3	Deflection of a Nonhomogeneous Beam Subjected to Axial Loading		261
		8.3.1	Thermostat Constructed from a Beam of Two Materials	263
		8.3.2	Deflection of Beam Constructed From Two Materials	263
		8.3.3	Program to Analyze Beam of Two Materials	264
		8.3.4	Program Output and Code	266
	8.4	Analysis of Pin-Connected Trusses		272
		8.4.1	Program to Analyze a Pin-Connected Truss	276
		8.4.2	Exercises	276
		8.4.3	Program Output and Code	277

A Utility Routines — 285

- A.1 A System Dependent Plot Save Function — 285
  - A.1.1 Program Output and Code — 285
- A.2 Polygonal Representation of a Circle — 286
  - A.2.1 Program Output and Code — 287
- A.3 Function to Flip Angle Measures — 287
  - A.3.1 Program Output and Code — 288
- A.4 Function for Polygon Clipping — 289
  - A.4.1 Equation of a Line — 289
  - A.4.2 Program to Clip Arbitrary Polygon — 290
  - A.4.3 Program Output and Code — 292

B Description of MATLAB Commands — 298

C List of MATLAB Routines with Descriptions — 309

Bibliography — 315

Index — 319

Preface

This book uses MATLAB® to solve various problems encountered in mechanics of materials. MATLAB[1] embodies an interactive environment for technical computing which includes a high level programming language and remarkably simple graphics commands facilitating two- and three-dimensional data presentation. The wealth of intrinsic functions to handle vector and matrix algebra greatly simplifies operations which would typically require numerous subroutines to accomplish. All the programs included in this book were developed using *The Student Edition of MATLAB* (version 4.x) and students should have a MATLAB manual readily available to use for reference.

This book is intended as a supplemental text for either a mechanics of materials course which incorporates the use of computers or a senior engineering analysis course. The goal was to utilize concepts learned in a first level engineering course to explore the benefits of computer usage in problem solution. It is anticipated that students will not have difficulty with the physics contained in these problems and will be able to spend the majority of their time learning how to exploit the use of computers for problem solving. Several important concepts related to solving engineering problems using a computer are presented in this text: a) the process of decomposing and structuring a problem so its solution can be efficiently implemented using a computer, b) the usefulness of graphics in the interpretation of results generated by computer analysis, c) the benefits of learning how to structure a problem to automatically generate the coefficients of a set of simultaneous equations (since MATLAB can solve these systems with an intrinsic operator), d) the use of simple iterative methods to solve problems, and e) the need to be aware of the potential for numerical errors which are a result of the numerical solution strategy. The authors wish to emphasize that the primary audience we address involves

---

[1]MATLAB is a registered trademark of The MathWorks, Inc. For additional information contact:

The MathWorks, Inc.
24 Prime Park Way
Natick, MA 01760-1500
(508) 647-7000, Fax: (508) 647-7001
Email: info@mathworks.com

people dealing with physical applications in mechanics. A thorough grounding in ideas of Euclidean geometry, mechanics of materials, and differential and integral calculus is essential to understand many of the topics.

Over twenty problems from mechanics of materials are investigated using MATLAB in this text. Topics discussed include axially loaded systems, stress-strain problems utilizing Hooke's law, statically indeterminate systems, the transformation of plane stress and plane strain, the transverse loading of cross sections, beam analysis using discontinuity functions, beams of several materials, and trusses. The generality of the solutions is enhanced significantly over methods presented in traditional texts by employing relationships which represent the geometrical properties of a cross section as a general polygon and an entire chapter is devoted to discussing the geometrical properties of polygons. Additionally, utilities to perform the task of "clipping" a polygon are included and are essential components of several of the general solutions for problems described by a polygonal cross section. Some of the examples are quite elementary. Others dealing with topics like indeterminate beams and trusses are more advanced. Finally, the majority of sections include student exercises intended to extend the student's experience with the material.

The programs utilize many of the vector and matrix operators available in MATLAB. However, the authors have not attempted to create the most "optimized" implementations. Since the objective is to illustrate the benefits of computers in solving engineering problems to students, clarity was always chosen over optimization. Additional optimization could be presented by the instructor or left as an assignment for students. The programs have extensive comments and are intended for study as separate entities without an additional reference. Consequently, some deliberate redundancy exists between program comments and text discussions. The authors have also used a program listing style which includes line numbers. Line numbers provide convenient reference points during discussions of particular program segments and are not inherent to MATLAB programs. The programs contain no interactive input for problem definition. The authors realize that interactive input increases the flexibility of the application but believe that the inclusion of all the data definitions within the program allows the student to readily understand the data entry requirements. Therefore, data definitions are contained within a logical `if` block at the beginning of each program. Data for different problem sets is selected by simply changing a single variable. Two additional characteristics of the programs in this book are reflective of publishing limitations. First, many of the programs utilize color graphics to improve clarity. However, all the graphics has been reproduced in black and white for this text. Second, the code listings are bound by page size restrictions and therefore contain many continuation lines. Finally, all of the programs are included on a diskette accompanying the book which is organized with directories corresponding to different chapters.

The majority of the problems contained in this text were originally developed as class assignments during the early 1980s. The completion of two computer assignments were required as part of a traditional three hour mechanics of materials course. The goal was to maintain and increase the computer fluency of students between an introductory programming language course and later engineering courses. The problems included in this text represent ideas suggested by several of our colleagues. The authors wish to acknowledge the contributions of Professors Samuel Gambrell and James Hill. Particularly, the authors would like to acknowledge the vision of Professor William Jordan, retired head of the Department of Engineering Mechanics at the University of Alabama, regarding the integration of personal computers into the engineering curriculum when such hardware first become readily available.

Louis H. Turcotte
turcotte@erc.msstate.edu

Howard B. Wilson
hwilson@ua1vm.ua.edu

Chapter 1

Axially Loaded Systems

This chapter presents two problems involving axially loaded members. The first problem consists of two axially loaded members supporting a gravity load. The problem demonstrates the value of an automatic solution when attempting to design members for minimum weight. The second problem uses an iterative method to determine the deflection in a two member truss. Both problems in this chapter require solving a set of simultaneous equations - a task performed easily using MATLAB. In Chapter 8, the concepts presented here are extended to analyze general trusses.

1.1 Analysis of Two Rods Subjected to Gravity Loading

Figure 1.1 depicts two rods attached to pin connections A and C. The rods support a gravity load P at point B. The distance between the supports, ℓ_{ac}, and the area of rod AB, A_{ab}, are considered constant. The densities of the material used for both rod AB, γ_{ab}, and rod BC, γ_{bc}, are known. The problem is to determine the total weight of the two rods for a range of angles defined by θ and a range of lengths of AB, ℓ_{ab}. The primary design requirements are that the stress in both rods must be equivalent and the area of BC, A_{bc}, is limited to a prescribed range of values.

1.1.1 Problem Solution

The solution of this problem requires that for any value of θ and ℓ_{ab} the forces in both rods be calculated so the stresses can be determined. Prior to writing the equations for static equilibrium the slope of rod BC must be determined. The law of cosines can be used to determine the length of BC when θ and ℓ_{ab} are known. Using Figure 1.2 as a reference the law of cosines can be written as

$$c^2 = a^2 + b^2 - 2ab\cos(\theta_C)$$

Therefore, the length of BC can be determined, or

$$\ell_{bc}^2 = \ell_{ac}^2 + \ell_{ab}^2 - 2\ell_{ac}\ell_{ab}\cos\theta$$

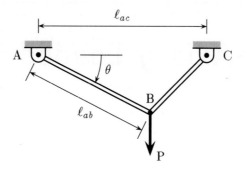

Figure 1.1. Problem Geometry

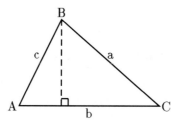

Figure 1.2. Law of Cosines Reference Triangle

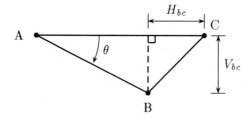

Figure 1.3. Dimensions for BC

1.1. Analysis of Two Rods Subjected to Gravity Loading

The slope of rod BC can now be determined using

$$V_{bc} = \ell_{ab} \sin\theta \qquad H_{bc} = \ell_{ac} - \ell_{ab} \cos\theta$$

as shown in Figure 1.3. The equations of static equilibrium provide two equations with two unknowns, or

$$\sum F_x = 0, \qquad \overbrace{(-\cos\theta)}^{a_{11}} F_{ab} + \overbrace{\left(\frac{H_{bc}}{\ell_{bc}}\right)}^{a_{12}} F_{bc} = 0$$

$$\sum F_y = 0, \qquad \overbrace{(\sin\theta)}^{a_{21}} F_{ab} + \overbrace{\left(\frac{V_{bc}}{\ell_{bc}}\right)}^{a_{22}} F_{bc} = P$$

The simultaneous solution of these two equations, using the coefficients a_{11}, a_{12}, a_{21}, and a_{22}, produces F_{ab} and F_{bc}. The stress in rod AB is

$$\sigma_{ab} = \frac{F_{ab}}{A_{ab}}$$

and the area of BC required to produce an equivalent stress is

$$A_{bc} = \frac{F_{bc}}{\sigma_{ab}}$$

Finally, the total weight in the two rods can be calculated, or

$$W = \gamma_{ab} A_{ab} \ell_{ab} + \gamma_{bc} A_{bc} \ell_{bc}$$

1.1.2 Program to Analyze Two Rods Subjected to Gravity Loading

The program written to solve this problem is summarized in Table 1.1. The results from the analysis are shown in several three-dimensional graphs.[1] These plots clearly show the variation of results as the two parameters θ and ℓ_{ab} are changed. Figure 1.4 depicts the variation of the force in rod AB. It is clear from this graph that small angles of θ create large forces in AB. Figure 1.5 indicates that the same is true for the forces in BC. The variation in the area in rod BC is depicted in Figures 1.6 and 1.7. Figure 1.7 presents only those solutions which satisfy the allowable range of areas for BC while Figure 1.6 shows all the solutions. The truncated plot is an effective method to identify the range of solutions. The variation in the total weight of rods AB and BC is shown in Figures 1.8 and 1.9. Figure 1.9 presents only

---

[1] The reader should be aware that these plots are rendered in color on your monitor. However, the graphs contained in this book are in black and white. Obviously, color greatly enhances the clarity of the plots.

those solutions which satisfy the allowable range of areas for BC while Figure 1.8 shows all the solutions. Once again the truncated plot reduces the effort required to interpret the results.

Table 1.1. Description of Code in Example **tworods**

Routine	Line	Operation
tworods		script file to execute program.
	38	select one of the example problems.
	39-45	define the input parameters.
	42	define the smallest and largest values of θ to consider. The number of angles to be analyzed is also defined.
	43	define the smallest and largest values of ℓ_{ab} to consider. The number of lengths to be analyzed is also defined.
	44	define the smallest and largest allowable values for the area of BC.
	48-49	create the vector of angles.
	50-51	create the vector of lengths.
	52-55	define some constants.
	58-77	loop on each angle of theta.
	60-76	loop on each length of AB.
	62-63	apply law of cosines.
	65-66	calculate rod BC dimensions.
	68-70	set up and solve simultaneous equations.
	71	store the results.
	73	calculate the stress in AB and the area of BC.
	74-75	calculate the weight of the rods.
	80-86	create 3D plot for the force in rod AB.
	89-95	create 3D plot for the force in rod BC.
	94	use intrinsic function **view** to set the viewing angle for the plot.
	98-104	create 3D plot for the total weight.
	103	save the axis settings using intrinsic function **axis** for later use.
		continued on next page

		continued from previous page
Routine	Line	Operation
	107-113	create 3D plot for the area of rod BC.
	112	save the axis settings using intrinsic function **axis** for later use.
	117-124	create information to only plot the weight and area of BC which satisfy the area of BC restrictions. Note this is accomplished by assigning the value of a point to the MATLAB constant **nan**. These points are not plotted by MATLAB's graphing routines.
	126-132	create 3D plot for the weights which are in the allowable range of areas for rod BC.
	135-141	create 3D plot for the areas of rod BC which are in the allowable range.
	131,140	force the axes for these plots to be identical with previous plots using the intrinsic **axis** function.

1.1.3 Exercises

1. Discuss the results shown in Figure 1.9. At what values of θ and ℓ_{ab} does the minimum weight occur? Provide an explanation of this result.

2. Investigate the results generated by program **tworods** as θ approaches angles of zero and ninety degrees.

3. Modify program **tworods** to provide the additional capability to specify an allowable range for stress (similar to the area range). Generate plots for stress, weight, and area which show only the allowable range results.

4. Add the capability to program **tworods** to determine the values for θ and ℓ_{AB} which will cause the weight to be a minimum within the allowable range of results.

5. Explore the use of the MATLAB plotting functions **contour**, **mesh**, **meshc**, **meshz**, **surfc**, and **waterfall** as a replacement for the **surf** routine used in **tworods**. Can you shift the values of the function to produce clearer plots of the combined information shown for some of these functions?

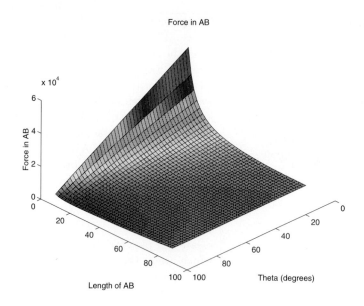

Figure 1.4. Force in Rod AB

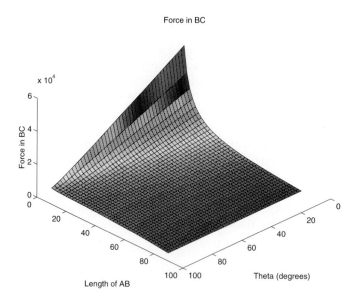

Figure 1.5. Force in Rod BC

1.1. Analysis of Two Rods Subjected to Gravity Loading

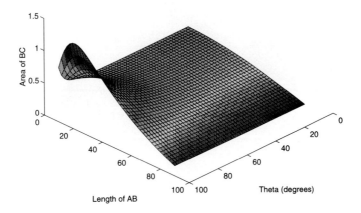

Figure 1.6. Area of Rod BC

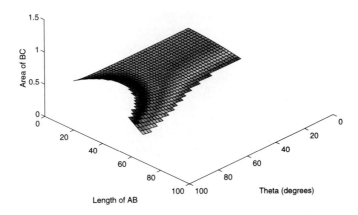

Figure 1.7. Area of Rod BC (Allowable Range)

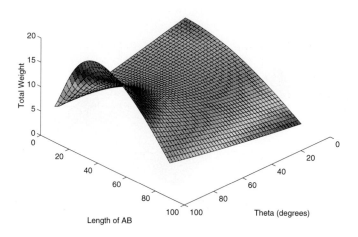

Figure 1.8. Total Weight of Both Rods

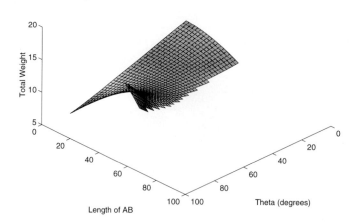

Figure 1.9. Total Weight of Both Rods (Allowable Range)

1.1. Analysis of Two Rods Subjected to Gravity Loading

1.1.4 Program Output and Code
Script File tworods

```
 1: % Example: tworods
 2: % ~~~~~~~~~~~~~~~~
 3: % This example analyzes the effects of a load
 4: % applied at the point of interconnect for
 5: % two rods.  The user defines a range of
 6: % lengths for the left rod (AB) and a range
 7: % of angles for rod AB measured clockwise from
 8: % the horizontal axis.
 9: %
10: % Data is defined in the declaration statements
11: % below, where:
12: %
13: % P            - downward vertical load
14: % L_ac         - horizontal span for AC
15: % A_ab         - area of rod AB
16: % Gamma_ab     - density of rod AB
17: % Gamma_bc     - density of rod BC
18: % Theta_range  - two element vector with the
19: %                starting and ending values
20: %                of theta to evaluate
21: %                (in degrees)
22: % No_thetas    - number of thetas to evaluate
23: % L_ab_range   - two element vector with the
24: %                starting and ending values
25: %                of length for AB to evaluate
26: % No_L_abs     - number of lengths of AB to
27: %                evaluate
28: % A_bc_range   - two element vector with the
29: %                minimum and maximum acceptable
30: %                values for the area of BC
31: %
32: % User m functions required:
33: %      genprint
34: %-----------------------------------------------
35:
36: clear;
37: %...Input definitions
38: Problem=1;
39: if Problem == 1
```

```
40:     P=5000; L_ac=100; A_ab=0.4;
41:     Gamma_ab=0.1; Gamma_bc=0.3;
42:     Theta_range=[5 85]; No_thetas=[20];
43:     L_ab_range=[5 95]; No_L_abs=[45];
44:     A_bc_range=[0.3 0.5];
45: end
46:
47: %...Set up some quantities
48: Theta=linspace(Theta_range(1), ...
49:               Theta_range(2),No_thetas+1);
50: L_ab=linspace(L_ab_range(1), ...
51:               L_ab_range(2),No_L_abs+1);
52: degrad=pi/180; thetai=Theta*degrad;
53: costheta=cos(thetai); sintheta=sin(thetai);
54: Lac2=L_ac*2; Lacsq=L_ac^2;
55: Lac2cos=Lac2*costheta; Labsq=L_ab.^2;
56:
57: %...Loop on each theta
58: for i=1:No_thetas+1
59:    %...Loop on each length
60:    for j=1:No_L_abs+1
61:       %...Law of cosines
62:       L_bc(i,j)=sqrt(Labsq(j)+Lacsq- ...
63:              L_ab(j)*Lac2cos(i));
64:       %...Rod BC slope
65:       Vbc=L_ab(j)*sintheta(i);
66:       Hbc=L_ac-L_ab(j)*costheta(i);
67:       %...Set up simultaneous equations and solve
68:       a=[-costheta(i)  Hbc/L_bc(i,j); ...
69:           sintheta(i)  Vbc/L_bc(i,j)];
70:       b=[0;P]; F=a\b;
71:       F_ab(i,j)=F(1); F_bc(i,j)=F(2);
72:       %...Calculate stress and weight
73:       Sigma=F(1)/A_ab; A_bc(i,j)=F(2)/Sigma;
74:       W(i,j)=Gamma_ab*A_ab*L_ab(j)+ ...
75:              Gamma_bc*A_bc(i,j)*L_bc(i,j);
76:    end
77: end
78:
79: %...Plot results
80: clf; surf(L_ab,Theta,F_ab);
81:    title('Force in AB');
82:    xlabel('Theta (degrees)');
83:    ylabel('Length of AB');
```

1.1. Analysis of Two Rods Subjected to Gravity Loading 11

```
84:    zlabel('Force in AB');
85:    view(135.0,45); drawnow;
86: % genprint('Fab');
87:    disp('Press key to continue'); pause;
88:
89: clf; surf(L_ab,Theta,F_bc)
90:    title('Force in BC');
91:    xlabel('Theta (degrees)');
92:    ylabel('Length of AB');
93:    zlabel('Force in BC');
94:    view(135.0,45); drawnow;
95: % genprint('Fbc');
96:    disp('Press key to continue'); pause;
97:
98: clf; surf(L_ab,Theta,W);
99:    title('Total Weight');
100:   xlabel('Theta (degrees)');
101:   ylabel('Length of AB');
102:   zlabel('Total Weight');
103:   view(135.0,45); drawnow; a1=axis;
104: % genprint('W');
105:   disp('Press key to continue'); pause;
106:
107: clf; surf(L_ab,Theta,A_bc);
108:   title('Area of BC');
109:   xlabel('Theta (degrees)');
110:   ylabel('Length of AB');
111:   zlabel('Area of BC');
112:   view(135.0,45); drawnow; a2=axis;
113: % genprint('Abc');
114:   disp('Press key to continue'); pause;
115:
116: %...Choose only prescribed range of Abc
117: for i=1:No_thetas+1
118:   for j=1:No_L_abs+1
119:     if A_bc(i,j) < A_bc_range(1) | ...
120:                   A_bc(i,j) > A_bc_range(2)
121:       W(i,j)=nan; A_bc(i,j)=nan;
122:     end
123:   end
124: end
125:
126: clf; surf(L_ab,Theta,W);
127:   title('Weight (allowable range)');
```

```
128:    xlabel('Theta (degrees)');
129:    ylabel('Length of AB');
130:    zlabel('Total Weight');
131:    view(135.0,45); axis([a1]); drawnow;
132: % genprint('Wallow');
133:    disp('Press key to continue'); pause;
134:
135: clf; surf(L_ab,Theta,A_bc);
136:    title('Area of BC (allowable range)');
137:    xlabel('Theta (degrees)');
138:    ylabel('Length of AB');
139:    zlabel('Area of BC');
140:    view(135.0,45); axis([a2]); drawnow;
141: % genprint('Abcallow');
```

1.2 Analysis of a Two Member Truss

Figure 1.10 shows a two member truss attached to a wall[2] at points A and C. The truss is subjected to a gravity load, P, at point B and will undergo deformation resulting in point B having a new position. The analysis of this configuration poses a common problem encountered in mechanics which can be addressed using an iterative solution methodolgy.

The forces determined from statics at point B cause an axial deformation in the truss members. After the deformation occurs point B will be at a new position and the angles θ_a, θ_b, and θ_c will no longer be the same as their original values. This means that the solution for the forces from statics is no longer valid since this solution utilized the original angles. However, the angles resulting after deformation can be used to recalculate the forces in the truss members using statics. After determining the new forces the deformation characteristics are recalculated. The procedure is repeated until the position of point B between two successive set of calculations does not change. This solution strategy is commonly referred to as an *iterative method*.

The problem shown in Figure 1.10 can be defined by specifying the truss member lengths (ℓ_{ab} and ℓ_{ac}), the gravity load applied at point B (P), the truss member areas (A_{ab} and A_{bc}), and the truss member moduli of elasticity (E_{ab} and E_{bc}).

1.2.1 Problem Solution

The first step towards analyzing the two member truss is to determine the initial values for the angles. For the original geometry the angle $(\theta_a)_o$ is 90°, therefore the length of member BC is

$$(\ell_{bc})_o = \sqrt{(\ell_{ab})_o^2 + (\ell_{ac})_o^2}$$

---

[2] Cook [7] has previously discussed the computer solution of this problem and implemented the solution in BASIC.

1.2. Analysis of a Two Member Truss

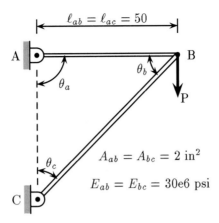

Figure 1.10. Problem Geometry

and the angles $(\theta_b)_o$ and $(\theta_c)_o$ are

$$(\theta_c)_o = \cos^{-1}\left[\frac{(\ell_{ac})_o}{(\ell_{bc})_o}\right] \qquad (\theta_b)_o = 90° - (\theta_c)_o$$

These lengths and angles will be used as the "starting" values for our iterative solution. Figure 1.11 diagrams the geometric parameters which will influence each iteration of the analysis. As point B changes position it will be necessary to utilize angles α and β to determine the components of force in each member in the x and y directions. Therefore, the values for lengths and angles can be defined for the first iteration, or

$$(\ell_{ab})_1 = (\ell_{ab})_o \qquad (\ell_{ac})_1 = (\ell_{ac})_o \qquad (\ell_{bc})_1 = (\ell_{bc})_o$$
$$(\theta_a)_1 = 90° \qquad (\theta_b)_1 = (\theta_b)_o \qquad (\theta_c)_1 = (\theta_c)_o$$
$$\alpha_1 = 0 \qquad \beta_1 = (\theta_b)_o$$

The forces in the two members can be determined by considering a free-body diagram at point B. This produces two equations which can be solved to determine the member forces, or

$$\sum F_x = 0, \qquad \overbrace{(-\cos\alpha_i)}^{a_{11}} F_{ab} + \overbrace{(\cos\beta_i)}^{a_{12}} F_{bc} = 0$$

$$\sum F_y = 0, \qquad \overbrace{(\sin\alpha_i)}^{a_{21}} F_{ab} + \overbrace{(\sin\beta_i)}^{a_{22}} F_{bc} = P$$

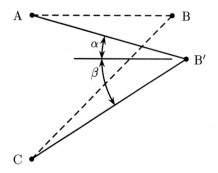

Figure 1.11. Deformed Geometry

where i represents the iteration number. The simultaneous solution of these two equations, using the coefficients a_{11}, a_{12}, a_{21}, and a_{22}, produces F_{ab} and F_{bc}. The deformation of the axially loaded truss members can be determined using

$$(\Delta_{ab})_i = \frac{(F_{ab})_i (\ell_{ab})_i}{A_{ab} E_{ab}} \qquad (\Delta_{bc})_i = \frac{-(F_{bc})_i (\ell_{bc})_i}{A_{bc} E_{bc}}$$

The deflections can be used to update the lengths and angles for the next iteration. The new lengths are

$$(\ell_{ab})_{i+1} = (\ell_{ab})_o + (\Delta_{ab})_i \qquad (\ell_{bc})_{i+1} = (\ell_{bc})_o + (\Delta_{bc})_i$$

The updated angles can be determined by recalling the following relationships for a general triangle[3] [39] as described by Figure 1.12.

$$s = \frac{1}{2}[(\ell_{ab})_i + (\ell_{bc})_i + (\ell_{ac})_o]$$

$$K = \sqrt{s[s - (\ell_{ab})_i][s - (\ell_{bc})_i][s - (\ell_{ac})_o]}$$

$$h_c = \frac{2K}{(\ell_{ac})_o}$$

Therefore, the updated values for the angles are

$$\alpha_{i+1} = \cos^{-1}\left[\frac{h_c}{(\ell_{ab})_i}\right] \qquad \beta_{i+1} = \cos^{-1}\left[\frac{h_c}{(\ell_{bc})_i}\right]$$

---

[3] The equation for K is known as Heron's formula.

1.2. Analysis of a Two Member Truss

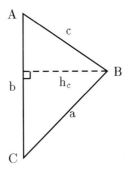

Figure 1.12. Reference Triangle

$$(\theta_a)_{i+1} = 90° - \alpha_{i+1} \qquad (\theta_b)_{i+1} = \alpha_{i+1} + \beta_{i+1}$$
$$(\theta_c)_{i+1} = 180° - (\theta_a)_{i+1} - (\theta_b)_{i+1}$$

The new position of point B can be determined, or

$$(\Delta_x)_i = (\ell_{ab})_i \cos(\alpha_{i+1}) - (\ell_{ab})_o$$
$$(\Delta_y)_i = (\ell_{bc})_i \sin(\beta_{i+1}) - (\ell_{ac})_o$$

This process can be continued until the change in the value of one of the control parameters such as α is essentially zero, or

$$|\alpha_i - \alpha_{i+1}| \approx 0$$

1.2.2 Program to Analyze a Two Member Truss

The program written to analyze the two member truss is summarized in Table 1.2. A partial output from the analysis is provided in Section 1.2.4. Figures 1.13 and 1.14 depict the progression of the iterative process for the analysis. Figure 1.14 also graphs the difference between successive values of α. Notice that this graph is plotted as a log function on the y-axis. It is easy to see from this plot how the convergence is proceeding. Finally, Figure 1.15 shows the position of point B as the iterative process occurs.

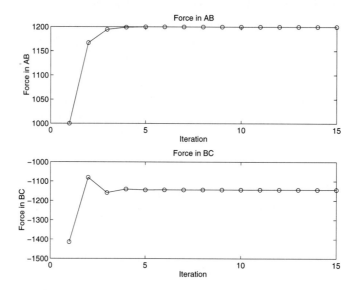

Figure 1.13. Forces in Truss Members

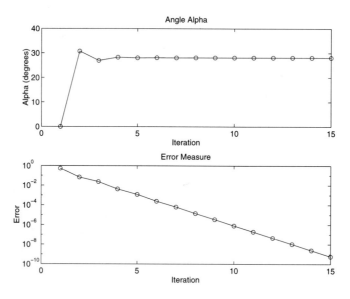

Figure 1.14. Error Analysis

1.2. Analysis of a Two Member Truss

Table 1.2. Description of Code in Example **truss**

Routine	Line	Operation
truss		script file to execute program.
	28	select one of the example problems.
	29-33	define the input parameters.
	32	specify the numerical definition for zero and specify a maximum number of iterations to attempt to achieve convergence.
	37-39	define the starting values.
	43	define the initial error value as a large quantity.
	45-68	output initial problem information.
	73-124	main iterative loop - note the use of `while` statement.
	75-78	keep program from looping indefinitely.
	80-82	set up and solve simultaneous equations.
	85-88	calculate new lengths.
	90-92	use triangle relationships to calculate h_c.
	94-98	calculate new angles.
	100-101	calculate new location of point B.
	102-103	calculate the difference between current solution and previous solution.
	105-123	output the results.
	128-162	plot the results.

1.2.3 Exercises

1. Modify the program **truss** to allow the value for the area of BC to vary. Have the program automatically determine the value for the area of BC which will produce

$$|\Delta_x| = r|\Delta_y|$$

   where r is the prescribed ratio to be obtained. Produce 3D plots for the results similar to those in **tworods**.

2. Modify the program **truss** to allow the value for the area of BC to vary. Have the program automatically determine the value for the area of BC which will produce

$$|F_{ab}| = r|F_{bc}|$$

   where r is the prescribed ratio to be obtained. Produce 3D plots for the results similar to those in **tworods**.

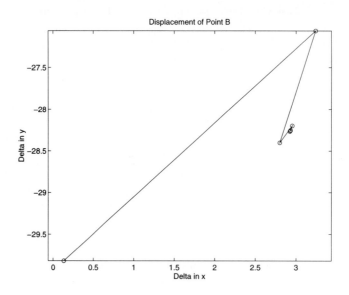

Figure 1.15. Displacement of Point B

3. Add the capability to generate plots for Δ_x versus iteration and Δ_y versus iteration.

4. Add the capability to plot the original truss geometry and deformed geometry on the same plot. Use solid lines for the original geometry and dashed lines for the deformed geometry.

1.2.4 Program Output and Code

Output from Example truss

```
Iterative Analysis of Truss
---------------------------

Original Properties
   Load (P):    1000
   Area AB:     2
   Area BC:     2
   E for AB:    3000
   E for BC:    3000
   Length AB:   50
   Length AC:   50
   Length BC:   70.7107
```

1.2. Analysis of a Two Member Truss

```
  Angle at A:   90 (degrees)
  Angle at B:   45 (degrees)
  Angle at C:   45 (degrees)
  Angle Alpha:  0 (degrees)
  Angle Beta:   45 (degrees)
  Epsilon:      1e-09
  max iter:     20

Iteration # 1
  Force AB :  1000
  Force BC : -1414.21
  Length AB: 58.3333
  Length BC: 54.044
  Delta AB:  0.166667
  Delta BC: -0.235702
  Delta x:   0.135136
  Delta y:  -29.8202
  Theta A:   59.2559 (degrees)
  Theta B:   52.6692 (degrees)
  Theta C:   68.0748 (degrees)
  Alpha:     30.7441 (degrees)
  Beta:      21.9252 (degrees)
  .
  . Material removed
  .

Iteration # 15
  Force AB :  1200
  Force BC : -1144.46
  Length AB: 60
  Length BC: 57.2231
  Delta AB:  0.2
  Delta BC: -0.190744
  Delta x:   2.93055
  Delta y:  -28.2552
  Theta A:   61.906 (degrees)
  Theta B:   50.4278 (degrees)
  Theta C:   67.6663 (degrees)
  Alpha:     28.094 (degrees)
  Beta:      22.3337 (degrees)
```

Script File truss

```
1:  % Example: truss
2:  % ~~~~~~~~~~~~~~
3:  % This example analyzes the effects of a load
4:  % applied to the point of interconnect for
5:  % two rods forming a truss.  The method
6:  % is based on iterating until the solution
7:  % reaches a prescribed convergence criteria.
8:  %
9:  % Data is defined in the declaration statements
10: % below, where:
11: %
12: % P        - downward vertical load
13: % L_ab_o   - original horizontal span for AB
14: % L_ac_o   - original horizontal span for AC
15: % A_ab     - area of rod AB
16: % A_bc     - area of rod AB
17: % E_ab     - modulus of elasticity for AB
18: % E_bc     - modulus of elasticity for AC
19: % eps      - solution tolerance
20: % max_iter - max # iterations
21: %
22: % User m functions required:
23: %      genprint
24: %-------------------------------------------------
25:
26: clear;
27: %...Input definitions
28: Problem=1;
29: if Problem == 1
30:    P=1000; L_ab_o=50; L_ac_o=50;
31:    A_ab=2.0; A_bc=2.0; E_ab=3000; E_bc=3000;
32:    eps=1.0e-9; max_iter=20;
33: end
34:
35: %...Initialize
36: raddeg=180/pi; pi2=pi/2;
37:
38: %...Calculate initial values
39: L_bc_o=sqrt(L_ab_o^2+L_ac_o^2);
40: Theta_a_o=pi2; Theta_c_o=acos(L_ac_o/L_bc_o);
41: Theta_b_o=pi2-Theta_c_o;
```

1.2. Analysis of a Two Member Truss

```
42:
43: %...Initialize loop parameters
44: Alpha(1)=0; Beta(1)=pi2-Theta_c_o;
45: error=10^10; i=0;
46:
47: fprintf('\n\nIterative Analysis of Truss');
48: fprintf(   '\n--------------------------');
49: fprintf('\n\nOriginal Properties');
50: fprintf(   '\n   Load (P):     %g',P);
51: fprintf(   '\n   Area AB:      %g',A_ab);
52: fprintf(   '\n   Area BC:      %g',A_bc);
53: fprintf(   '\n   E for AB:     %g',E_ab);
54: fprintf(   '\n   E for BC:     %g',E_bc);
55: fprintf(   '\n   Length AB:    %g',L_ab_o);
56: fprintf(   '\n   Length AC:    %g',L_ac_o);
57: fprintf(   '\n   Length BC:    %g',L_bc_o);
58: fprintf('   \n   Angle at A:   %g (degrees)', ...
59:    Theta_a_o*raddeg);
60: fprintf('   \n   Angle at B:   %g (degrees)', ...
61:    Theta_b_o*raddeg);
62: fprintf('   \n   Angle at C:   %g (degrees)', ...
63:    Theta_c_o*raddeg);
64: fprintf('   \n   Angle Alpha:  %g (degrees)', ...
65:    Alpha(1)*raddeg);
66: fprintf('   \n   Angle Beta:   %g (degrees)', ...
67:    Beta(1)*raddeg);
68: fprintf(   '\n   Epsilon:      %g',eps);
69: fprintf(   '\n   max iter:     %g',max_iter);
70: fprintf('\n');
71:
72: %.................................
73: %...Loop until convergence achieved
74: %.................................
75: while error>eps
76:    i=i+1;
77:    if i>max_iter
78:       fprintf('\n\n Max iterations exceeded\n');
79:       error('\nProgram aborted');
80:    end
81:    %...Set up simultaneous equations and solve
82:    a=[-cos(Alpha(i)) cos(Beta(i)); ...
83:        sin(Alpha(i)) sin(Beta(i))];
84:    b=[0;P]; F=a\b;
85:    F_ab(i)=F(1); F_bc(i)=F(2);
```

```
86:     %...Calculate new lengths
87:     delta_ab(i)=F_ab(i)/(A_ab*E_ab);
88:     delta_bc(i)=-F_bc(i)/(A_bc*E_bc);
89:     L_ab(i)=L_ab_o*(1+delta_ab(i));
90:     L_bc(i)=L_bc_o*(1+delta_bc(i));
91:     %...Use triangle definitions
92:     S=0.5*(L_ab(i)+L_bc(i)+L_ac_o);
93:     K=sqrt(S*(S-L_ab(i))*(S-L_bc(i))*(S-L_ac_o));
94:     Hc=2*K/L_ac_o;
95:     %...New angles
96:     Alpha(i+1)=acos(Hc/L_ab(i));
97:     Beta(i+1)=acos(Hc/L_bc(i));
98:     Theta_a(i)=pi2-Alpha(i+1);
99:     Theta_b(i)=Alpha(i+1)+Beta(i+1);
100:    Theta_c(i)=pi-Theta_a(i)-Theta_b(i);
101:    %...New location of pin B
102:    delta_x(i)=L_ab(i)*cos(Alpha(i+1))-L_ab_o;
103:    delta_y(i)=-(L_ac_o-L_bc(i)*sin(Beta(i+1)));
104:    error=abs(Alpha(i)-Alpha(i+1));
105:    err(i)=error;
106:
107:    fprintf('\n\nIteration # %g',i);
108:    fprintf('\n   Force AB : %g',F_ab(i));
109:    fprintf('\n   Force BC : %g',-F_bc(i));
110:    fprintf('\n   Length AB: %g',L_ab(i));
111:    fprintf('\n   Length BC: %g',L_bc(i));
112:    fprintf('\n   Delta AB:  %g',delta_ab(i));
113:    fprintf('\n   Delta BC:  %g',delta_bc(i));
114:    fprintf('\n   Delta x:   %g',delta_x(i));
115:    fprintf('\n   Delta y:   %g',delta_y(i));
116:    fprintf('\n   Theta A:   %g (degrees)', ...
117:            Theta_a(i)*raddeg);
118:    fprintf('\n   Theta B:   %g (degrees)', ...
119:            Theta_b(i)*raddeg);
120:    fprintf('\n   Theta C:   %g (degrees)', ...
121:            Theta_c(i)*raddeg);
122:    fprintf('\n   Alpha:     %g (degrees)', ...
123:            Alpha(i+1)*raddeg);
124:    fprintf('\n   Beta:      %g (degrees)', ...
125:            Beta(i+1)*raddeg);
126: end
127: fprintf('\n\n');
128:
129: %...Plot results
```

1.2. Analysis of a Two Member Truss

```
130:  clf; iter=1:i; F_bc=-F_bc;
131:  subplot(2,1,1);
132:    plot(iter,F_ab(1:i),'o',iter,F_ab(1:i),'-')
133:    title('Force in AB');
134:    xlabel('Iteration'); ylabel('Force in AB');
135:  subplot(2,1,2);
136:    plot(iter,F_bc(1:i),'o',iter,F_bc(1:i),'-')
137:    title('Force in BC');
138:    xlabel('Iteration'); ylabel('Force in BC');
139:    drawnow;
140:  % genprint('force');
141:    disp('Press key to continue'); pause;
142:
143:  clf; Alpha=Alpha*raddeg;
144:  subplot(2,1,1);
145:    plot(iter,Alpha(1:i),'o', ...
146:         iter,Alpha(1:i),'-')
147:    title('Angle Alpha'); xlabel('Iteration');
148:    ylabel('Alpha (degrees)');
149:  subplot(2,1,2);
150:    semilogy(iter,err(1:i),'o',iter,err(1:i),'-')
151:    title('Error Measure');
152:    xlabel('Iteration'); ylabel('Error');
153:    drawnow;
154:  % genprint('error');
155:    disp('Press key to continue'); pause;
156:
157:  clf;
158:  plot(delta_x(1:i),delta_y(1:i),'o', ...
159:       delta_x(1:i),delta_y(1:i),'-')
160:    axis('equal');
161:    title('Displacement of Point B');
162:    xlabel('Delta in x'); ylabel('Delta in y');
163:    drawnow;
164:  % genprint('deltab');
```

Chapter 2

Stress and Strain

This chapter presents two straight forward problems involving stress and strain. The first problem determines the stresses, or strains, and related properties for a plate obeying Hooke's Law. This problem uses a short computer program to automate the repetitious evaluation of equations involving several terms. The second problem illustrates how a simple computer program can be used to automate the determination of shear stresses in bolts used in a flange coupling. This problem demonstrates how effortlessly the coefficients for a set of simultaneous equations can be set up and how simple it is to solve this set of equations using a computer.

2.1 Plate Analysis Using Hooke's Law

A plate, shown in Figure 2.1, can be loaded by stresses in the x, y, and z directions. According to Hooke's Law for a generalized three-dimensional plate the following relationships for strain in terms of stress can be defined.

$$\epsilon_x = \frac{1}{E}[\sigma_x - \mu(\sigma_y + \sigma_z)]$$

$$\epsilon_y = \frac{1}{E}[\sigma_y - \mu(\sigma_x + \sigma_z)]$$

$$\epsilon_z = \frac{1}{E}[\sigma_z - \mu(\sigma_x + \sigma_y)]$$

These equations can also be written for stresses in terms of strains, or

$$\sigma_x = \frac{E}{(1+\mu)(1-2\mu)}[(1-\mu)\epsilon_x + \mu\epsilon_y + \mu\epsilon_z]$$

$$\sigma_y = \frac{E}{(1+\mu)(1-2\mu)}[(1-\mu)\epsilon_y + \mu\epsilon_x + \mu\epsilon_z]$$

$$\sigma_z = \frac{E}{(1+\mu)(1-2\mu)}[(1-\mu)\epsilon_z + \mu\epsilon_x + \mu\epsilon_y]$$

2.1. Plate Analysis Using Hooke's Law

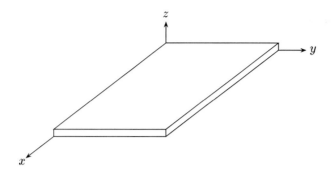

Figure 2.1. Plate Geometry

where E is the modulus of elasticity and μ is Poisson's ratio. A third material constant, called the modulus of rigidity is defined as

$$G = \frac{E}{2(1+\mu)}$$

Therefore, given any two of the three material constants the remaining material constant can be calculated, or

$$E = 2G(1+\mu) \qquad \mu = \frac{E}{2G} - 1$$

For the above equations positive strains can be defined as those indicating expansion while negative strains indicate contraction. Likewise, positive stresses are those which induce tension and negative stresses are those which induce compression. One additional quantity of interest is defined as the *dilatation*, or *volumetric strain*. Dilatation is a measure of the change in volume per unit volume, or

$$e = \frac{1 - 2\mu}{E}(\sigma_x + \sigma_y + \sigma_z)$$

2.1.1 Program to Calculate Stresses or Strains in a Plate

Program **hooke** was written to solve for the stresses or strains in a plate obeying Hooke's Law. Section 2.1.3 contains the output from the analysis of a plate. The program calculates the stresses, or strains, based on which is defined in the problem specification. The final lengths, areas, and volume are calculated along with the change in lengths, areas, and volume. The program is summarized in Table 2.1.

Routine	Line	Operation
hooke		script file to execute program.
	30	select one of the example problems.
	31-46	define the input parameters.
	33	μ is set to zero, therefore calculate it from E and G.
	35	stress array is set to empty vector which indicates that stresses are to be calculated.
	48-55	calculate one of constants if necessary.
	59-65	calculate strains from stresses.
	67-74	calculate stresses from strains.
	78	calculate change in lengths.
	80-89	calculate original and final area of plate.
	91-93	calculate original and final volume of plate.
	96	calculate dilatation.
	98-139	output results.

Table 2.1. Description of Code in Example **hooke**

2.1.2 Exercises

1. Add the capability to routine **hooke** to create various plots of the plate's deformation. For example, a plot of the $x - y$ plane showing both the original geometry and the deformed geometry. (The deformed geometry will most likely require the exaggeration of the deformation to make it noticeable.)

2.1.3 Program Output and Code

Output from Example hooke

```
Application of Hooke's Law to Plates
------------------------------------

Constants:
  Modulus of Elasticity: 5e+10
  Modulus of Rigidity:   1.875e+10
  Poisson's ratio:       0.333333

Stresses:
  Sigma-x: -1.8e+08
  Sigma-y: -6e+07
  Sigma-z: 0
```

2.1. Plate Analysis Using Hooke's Law

```
Strains:
  Epsilon-x: -0.0032
  Epsilon-y: 0
  Epsilon-z: 0.0016

Lengths:
  Lx (original): 0.1
  Ly (original): 0.04
  Lz (original): 0.025
  Lx (final):    0.09968
  Ly (final):    0.04
  Lz (final):    0.02504
  Lx (delta):    -0.00032
  Ly (delta):    0
  Lz (delta):    4e-05

Areas:
  A-xy (original): 0.004
  A-xz (original): 0.0025
  A-yz (original): 0.001
  A-xy (final):    0.0039872
  A-xz (final):    0.00249599
  A-yz (final):    0.0010016
  A-xy (delta):    -1.28e-05
  A-xz (delta):    -4.0128e-06
  A-yz (delta):    1.6e-06

Volumes:
  Original: 0.0001
  Final:    9.98395e-05
  Delta:    -1.60512e-07

Dilatation: -0.0016
```

Script File hooke

```
1: % Example: hooke
2: % ~~~~~~~~~~~~~~~
3: % This program calculates the stresses or
4: % strains in a plate using Hooke's law.
5: %
6: % Data is defined in the declaration statements
7: % below, where:
```

```
%
% Emod      - modulus of elasticity
% mu        - Poisson's ratio
% ModRigid  - modulus of rigidity
% Length    - vector containing Lx, Ly, Lz
% Stress    - vector containing stresses
% Strain    - vector containing strains
%
% NOTES: 1) for Emod, mu, ModRigid only two
%           values are required.  Set the
%           unknown one to zero and the program
%           will calculate this value.
%        2) if stresses are provided, set the
%           Strain input to empty vector. If
%           strains are provided, set the
%           Stress input to empty vector.
%
% User m functions required: none
%-----------------------------------------------

clear;
%...Input definitions
Problem=3;
if Problem == 1
   %...Strain, E, G given
   Emod=30e6; mu=0; ModRigid=12e6;
   Length=[96 144 0.25];
   Stress=[]; Strain=[-.0008 -.0006 0];
elseif Problem == 2
   %...Stress, E, mu given
   Emod=2e11; mu=0.3; ModRigid=0;
   Length=[1e-1 6e-3 7.5e-2];
   Stress=[1.5e8 0 1e8]; Strain=[];
elseif Problem == 3
   %...Stress, G, mu given
   Emod=0; mu=1/3; ModRigid=18.75e9;
   Length=[0.1 0.04 0.025];
   Stress=[-1.8e8 -6e7 0]; Strain=[];
end

%...Calculate if necessary
if Emod == 0
   Emod=2*ModRigid*(1+mu);
elseif mu == 0
```

2.1. Plate Analysis Using Hooke's Law

```
52:     mu=(Emod/(2*ModRigid))-1;
53: elseif ModRigid == 0
54:     ModRigid=Emod/(2*(1+mu));
55: end
56:
57: %...Calculate stresses or strains
58: if length(Stress) ~= 0
59:     %...Find the strains
60:     Strain(1)=1/Emod* ...
61:        (Stress(1)-mu*(Stress(2)+Stress(3)));
62:     Strain(2)=1/Emod* ...
63:        (Stress(2)-mu*(Stress(1)+Stress(3)));
64:     Strain(3)=1/Emod* ...
65:        (Stress(3)-mu*(Stress(1)+Stress(2)));
66: elseif length(Strain) ~= 0
67:     %...Find the stresses
68:     fact=Emod/((1+mu)*(1-2*mu));
69:     Stress(1)=fact*((1-mu)*Strain(1)+ ...
70:             mu*Strain(2)+mu*Strain(3));
71:     Stress(2)=fact*((1-mu)*Strain(2)+ ...
72:             mu*Strain(1)+mu*Strain(3));
73:     Stress(3)=fact*((1-mu)*Strain(3)+ ...
74:             mu*Strain(1)+mu*Strain(2));
75: end
76:
77: %...Change in lengths and final lengths
78: Delta=Strain.*Length; FLength=Length+Delta;
79:
80: %...Area calculations
81: Area_xy=Length(1)*Length(2);
82: Area_xz=Length(1)*Length(3);
83: Area_yz=Length(2)*Length(3);
84: FArea_xy=FLength(1)*FLength(2);
85: FArea_xz=FLength(1)*FLength(3);
86: FArea_yz=FLength(2)*FLength(3);
87: DArea_xy=FArea_xy-Area_xy;
88: DArea_xz=FArea_xz-Area_xz;
89: DArea_yz=FArea_yz-Area_yz;
90:
91: %...Volume calculations
92: Volume=prod(Length); FVolume=prod(FLength);
93: DVolume=FVolume-Volume;
94:
95: %...Dilatation
```

```
 96: Dilat=(1-2*mu)/Emod*sum(Stress);
 97:
 98: fprintf( ...
 99:    '\n\nApplication of Hooke''s Law to Plates');
100: fprintf( ...
101:    '\n------------------------------------');
102: fprintf('\n\nConstants:');
103: fprintf('\n  Modulus of Elasticity: %g', Emod);
104: fprintf('\n  Modulus of Rigidity:   %g', ...
105:         ModRigid);
106: fprintf('\n  Poisson''s ratio:       %g', mu);
107: fprintf('\n\nStresses:');
108: fprintf( '\n  Sigma-x: %g',Stress(1));
109: fprintf( '\n  Sigma-y: %g',Stress(2));
110: fprintf( '\n  Sigma-z: %g',Stress(3));
111: fprintf('\n\nStrains:');
112: fprintf( '\n  Epsilon-x: %g',Strain(1));
113: fprintf( '\n  Epsilon-y: %g',Strain(2));
114: fprintf( '\n  Epsilon-z: %g',Strain(3));
115: fprintf('\n\nLengths:');
116: fprintf( '\n  Lx (original): %g',Length(1));
117: fprintf( '\n  Ly (original): %g',Length(2));
118: fprintf( '\n  Lz (original): %g',Length(3));
119: fprintf( '\n  Lx (final):    %g',FLength(1));
120: fprintf( '\n  Ly (final):    %g',FLength(2));
121: fprintf( '\n  Lz (final):    %g',FLength(3));
122: fprintf( '\n  Lx (delta):    %g',Delta(1));
123: fprintf( '\n  Ly (delta):    %g',Delta(2));
124: fprintf( '\n  Lz (delta):    %g',Delta(3));
125: fprintf('\n\nAreas:');
126: fprintf( '\n  A-xy (original): %g',Area_xy);
127: fprintf( '\n  A-xz (original): %g',Area_xz);
128: fprintf( '\n  A-yz (original): %g',Area_yz);
129: fprintf( '\n  A-xy (final):    %g',FArea_xy);
130: fprintf( '\n  A-xz (final):    %g',FArea_xz);
131: fprintf( '\n  A-yz (final):    %g',FArea_yz);
132: fprintf( '\n  A-xy (delta):    %g',DArea_xy);
133: fprintf( '\n  A-xz (delta):    %g',DArea_xz);
134: fprintf( '\n  A-yz (delta):    %g',DArea_yz);
135: fprintf('\n\nVolumes:');
136: fprintf( '\n  Original: %g',Volume);
137: fprintf( '\n  Final:    %g',FVolume);
138: fprintf( '\n  Delta:    %g',DVolume);
139: fprintf('\n\nDilatation: %g \n',Dilat);
```

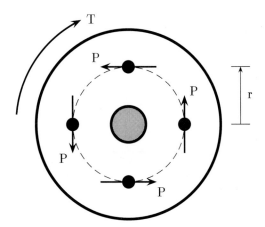

Figure 2.2. Flange Coupling - Single Ring of Bolts

2.2 Stresses in the Bolts of a Flange Coupling

Shafts are commonly joined end-to-end using a flange coupling. The flange coupling transfers the torque applied to one shaft to the other shaft via flange bolts (or rivets). The design of the coupling is dependent on the ability to determine the stresses induced on the cross section of each bolt. This is necessary to assure that the bolt's allowable stresses are not exceeded. Figure 2.2 shows the relationship between the applied torque and the bolt forces necessary to transfer the torque. For equilibrium to be satisfied the following relationship must be true

$$T = nrP$$

where T is the total torque applied to the shaft, n is the number of bolts, r is the radius arm to the bolt circle, and P is the total force transferred by a single bolt. The shearing stress on the cross section of a single bolt is

$$\tau = \frac{P}{A}$$

where A is the cross-sectional area of a single bolt. The shear stress can be substituted into the torque equation to yield

$$T = nr\tau A$$

Often the bolts of a flange coupling are arranged in more than one concentric circle as shown in Figure 2.3. When more than one circle of bolts is employed the

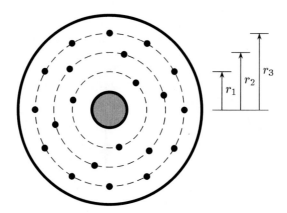

Figure 2.3. Flange Coupling - Multiple Rings of Bolts

shearing stresses are assumed to vary linearly with the radii of the bolt circles, or for the flange shown in Figure 2.3 the relationship is

$$\frac{\tau_1}{r_1} = \frac{\tau_2}{r_2} = \frac{\tau_3}{r_3}$$

and the torque is

$$T = n_1 r_1 \tau_1 A_1 + n_2 r_2 \tau_2 A_2 + n_3 r_3 \tau_3 A_3$$

2.2.1 Program to Determine Bolt Stresses in a Flange Coupling

From the equations presented in the previous section it is easy to craft a general solution to the problem of computing the bolt stresses in a flange coupling for a specified configuration. The torque equation can be written as:

$$T = \sum_{i=1}^{k} n_i r_i \tau_i A_i$$

where k is the number of concentric bolt circles, n is the number of bolts in a circle, r is the radius arm to a bolt circle, τ is the shear stress for a single bolt in a bolt circle, and A is the area of a single bolt in a bolt circle. This single equation has k unknown stresses. Therefore, $k-1$ additional equations are necessary to solve for the stresses.

The linear shear stress relationship can be written as

$$\frac{\tau_i}{r_i} = \frac{\tau_{i+1}}{r_{i+1}} \qquad i = 1, \ldots, k-1$$

2.2. Stresses in the Bolts of a Flange Coupling

and can be rearranged to produce

$$r_{i+1}\tau_i - r_i\tau_{i+1} = 0 \qquad i = 1, \ldots, k-1$$

This provides an additional $k-1$ equations and combined with the equation for torque the system can be solved for the k unknown bolt shear stresses. This set of simultaneous equations has the form

$$\begin{aligned}
a_{11}\tau_1 &+ a_{12}\tau_2 + \cdots + a_{1k}\tau_k = b_1 \\
a_{21}\tau_1 &+ a_{22}\tau_2 + \cdots + a_{2k}\tau_k = b_2 \\
&\vdots \\
a_{k1}\tau_1 &+ a_{k2}\tau_2 + \cdots + a_{kk}\tau_k = b_k
\end{aligned}$$

The linear stress relationship produces a set of coefficients on the matrix diagonal defined by

$$a_{ij} = r_{i+1} \quad \text{when} \quad i = j \qquad i = 1, \ldots, k-1$$

and a set of coefficients for the off-diagonal terms which are defined by

$$a_{ij} = -r_i \quad \text{when} \quad i = j-1 \qquad i = 1, \ldots, k$$

The remaining off-diagonal terms are zero. The torque equation yields a single set of coefficients, or

$$a_{ki} = n_i r_i A_i \qquad i = 1, \ldots, k$$

The values for b are defined by

$$b_i = 0 \qquad i = 1, \ldots, k-1$$

and

$$b_k = T$$

With all the coefficients defined the set of simultaneous equations can be solved for the unknown bolt stresses.

Program **flbolts** was written to solve for the shearing stresses in a flange coupling and is summarized in Table 2.2. Figure 2.4 depicts the geometry of the sample dataset and Section 2.2.3 contains the output from the analysis of this dataset. The reader should notice the inclusion of an equilibrium check after the analysis is complete. This type of calculation is often effectively used to catch basic programming mistakes and problem definition errors.

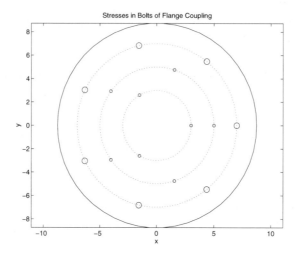

Figure 2.4. Geometry for Bolts of Flange Coupling

Routine	Line	Operation
flbolts		script file to execute program.
	22	select one of the example problems.
	23-28	define the input parameters.
	30-31	zero the the matrix of coefficients and vector of right-hand-side values.
	37-46	define the coefficients produced by the linear stress requirement.
	48-51	define the coefficients produced by the torque equation.
	53	solve the set of simultaneous equations.
	60-82	output the results.
	86-87	create a shaft geometry for plotting.
	87	use function **circle** to create points on circle. (See Appendix A.)
	89-108	create the plot of the geometry.

Table 2.2. Description of Code in Example **flbolts**

2.2. Stresses in the Bolts of a Flange Coupling

2.2.2 Exercises

1. Add the capability to plot the stress versus the bolt radius. What is the nature of this distribution?

2. Add the capability to plot the total force carried by all the bolts at each radius arm versus the the radius arm for the bolts. Is the relationship linear?

3. Add the capability to plot the total torque carried by all the bolts at each radius arm versus the the radius arm for the bolts. What can you deduce concerning which ring of bolts best carries the torque?

4. Modify the program **flbolts** to allow bolts at the same radius arm to have different diameters.

2.2.3 Program Output and Code

Output from Example flbolts

```
Bolt Stresses for a Flange Coupling
-----------------------------------

Bolt circle: 1
   Number of bolts:             3
   Bolt diameter:               0.25
   Radius arm for bolt circle:  3
   Stress:                      4010.2

Bolt circle: 2
   Number of bolts:             5
   Bolt diameter:               0.25
   Radius arm for bolt circle:  5
   Stress:                      6683.67

Bolt circle: 3
   Number of bolts:             7
   Bolt diameter:               0.5
   Radius arm for bolt circle:  7
   Stress:                      9357.14

Equilibrium check
   Applied Torque:   100000
   Computed Torque:  100000
```

Script File flbolts

```
 1: % Example: flbolts
 2: % ~~~~~~~~~~~~~~~~
 3: % This example determines the stresses
 4: % in the bolts of a flange coupling.
 5: %
 6: % Data is defined in the declaration statements
 7: % below, where:
 8: %
 9: % Torque            - applied torque
10: % Bolt_diameter     - vector of bolt
11: %                     diameters at each radius
12: % Radius_arm_bolts  - vector of radius arms
13: % No_bolts_at_radius - vector with the number
14: %                     of bolts at each radius
15: %
16: % User m functions required:
17: %     circle, genprint
18: %-----------------------------------------
19:
20: clear;
21: %...Input definitions
22: Problem=1;
23: if Problem == 1
24:    Torque=1e5;
25:    Bolt_diameter     =[0.25 0.25 0.5];
26:    Radius_arm_bolts  =[3    5    7   ];
27:    No_bolts_at_radius=[3    5    7   ];
28: end
29: No_circles=length(No_bolts_at_radius);
30: pi4=pi/4; b=zeros(No_circles,1);
31: A=zeros(No_circles,No_circles);
32:
33: %.........................................
34: %...Set up & solve simultaneous equations
35: %.........................................
36: %...Coefficients from linear stress requirement
37: for i=1:No_circles-1
38:    for j=1:No_circles
39:       if i == j
40:          A(i,j)=Radius_arm_bolts(i+1);
41:       end
```

2.2. Stresses in the Bolts of a Flange Coupling

```
42:      if i == j-1
43:        A(i,j)= -Radius_arm_bolts(i);
44:      end
45:    end
46: end
47: %...Coefficients from statics
48: b(No_circles)=abs(Torque);
49: A(No_circles,:)=No_bolts_at_radius.* ...
50:                 (pi4*Radius_arm_bolts).* ...
51:                 Bolt_diameter.^2;
52: %...Solve
53: Tau=A\b; Tau=Tau';
54:
55: %...Verify equilibrium
56: Check=sum(No_bolts_at_radius.* ...
57:           (pi4*Radius_arm_bolts).* ...
58:           Bolt_diameter.^2.*Tau);
59:
60: fprintf( ...
61:   '\n\nBolt Stresses for a Flange Coupling');
62: fprintf( ...
63:   '\n--------------------------------');
64: for i=1:No_circles
65:   fprintf('\n\nBolt circle: %g',i);
66:   fprintf( ...
67:     '\n  Number of bolts:            %g', ...
68:     No_bolts_at_radius(i));
69:   fprintf( ...
70:     '\n  Bolt diameter:              %g', ...
71:     Bolt_diameter(i));
72:   fprintf( ...
73:     '\n  Radius arm for bolt circle: %g', ...
74:     Radius_arm_bolts(i));
75:   fprintf( ...
76:     '\n  Stress:                     %g', ...
77:     Tau(i));
78: end
79: fprintf('\n\nEquilibrium check');
80: fprintf('\n  Applied Torque:  %g',Torque);
81: fprintf('\n  Computed Torque: %g',Check);
82: fprintf('\n');
83:
84: %...Assume shaft is a bit larger than last
85: %...bolt ring
```

```
86:  radius=1.25*Radius_arm_bolts(No_circles);
87:  [xs,ys]=circle(50,0,0,radius);
88:
89:  clf;
90:  plot(xs,ys,'-'); hold on;
91:    for i=1:No_circles
92:      %...Bolt ring circles
93:      [x,y]=circle(50,0,0,Radius_arm_bolts(i));
94:      plot(x,y,':');
95:      %...Location of bolts in each ring
96:      [x,y]=circle(No_bolts_at_radius(i), ...
97:                   0,0,Radius_arm_bolts(i));
98:      %...Plot each bolt
99:      for j=1:No_bolts_at_radius(i)
100:       [xb,yb]=circle(20,x(j),y(j), ...
101:                      Bolt_diameter(i)/2);
102:       plot(xb,yb,'-');
103:      end
104:    end
105:    axis('equal');
106:    title('Stresses in Bolts of Flange Coupling');
107:    xlabel('x'); ylabel('y'); hold off; drawnow;
108: % genprint('flange');
```

Chapter 3

Statically Indeterminate Systems

Statically determinate systems represent a small fraction of the problems encountered in mechanics. These problems are characterized by solutions dependent on the application of simple equations of static equilibrium. Statically indeterminate systems cannot be solved by employing the equations of equilibrium alone. Additional relationships developed from the deformation characteristics of a problem are necessary to describe the problem. Indeterminate problems commonly require the solution of simultaneous systems of equations and are excellent candidates for computer implementation. This chapter includes computer solutions of three different indeterminate systems: a gravity load supported by multiple rods hanging from a ceiling, a rigid bar suspended by multiple cables from a ceiling, and an axially loaded shaft. Each problem demonstrates different solution methods and the corresponding computer implementation.

3.1 Multiple Rods with a Common Load Point

Figure 3.1 depicts a gravity load being carried by a set of rods suspended from a ceiling. The rods have different properties (area, modulus of elasticity, length, and attachment location/angle) and undergo a deflection due to the gravity loading W. Figure 3.2 shows the deflection which results from the load at the point where the rods are connected. The deflection of each axially loaded rod can be determined if the force in each rod is known. However, only two equations from statics are available and several rods could be used in the configuration. Additional relationships must be defined to determine the forces and the corresponding deflection at the load point. It should be readily apparent that additional equations may be developed using the fact that all the rods are connected at the load point and therefore must incur the same deflection. The combination of the equations of statics and the equations produced by this deflection requirement provide sufficient information to solve the system.

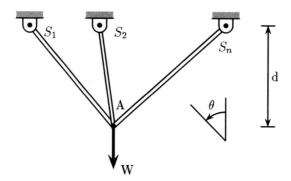

Figure 3.1. Problem Geometry

The equations for static equilibrium provide two relationships which can be written as

$$\sum F_x = 0, \qquad \sum_{i=1}^{n} F_i \sin\theta_i = 0$$

and

$$\sum F_y = 0, \qquad \sum_{i=1}^{n} F_i \cos\theta_i = W$$

where n is the number of rods, F_i is the force in a particular rod, and θ_i is the initial angle a rod forms with the vertical axis (θ is considered positive as defined in Figure 3.1). The deflection of each axially loaded member is defined by

$$\Delta_i = \frac{F_i L_i}{A_i E_i} \qquad \text{or} \qquad \Delta_i = f_i F_i \qquad i = 1,\ldots,n$$

where A is the area, E is the modulus of elasticity, L is the length, and f is called the *flexibility coefficient* of the rod. If the deflection of the system is considered small (the changes in value for θ following deformation do not adversely affect the solution) then a set of n equations relating the deflections can be written, or

$$f_i F_i - (\cos\theta_i)\Delta_y - (\sin\theta_i)\Delta_x = 0 \qquad i = 1,\ldots,n$$

This set of n equations has $n+2$ unknowns (the n forces, F_i, and two unknown deflections, Δ_x and Δ_y). However, the two equations from statics provide the two additional equations required to solve for the unknown forces and deflections. This set of $n+2$ equations can be written as a set of simultaneous equations with the

3.1. Multiple Rods with a Common Load Point

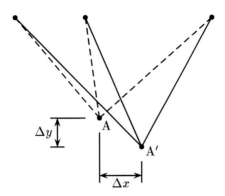

Figure 3.2. Deflected Geometry

form

$$
\begin{array}{ccccccccccc}
f_1 F_1 & + & 0 & + & \cdots & + & 0 & + & (-s_1)\Delta_x & + & (-c_1)\Delta_y & = & 0 \\
0 & + & f_2 F_2 & + & \cdots & + & 0 & + & (-s_2)\Delta_x & + & (-c_2)\Delta_y & = & 0 \\
\vdots & & \vdots & & \vdots & & \vdots & & \vdots & & \vdots & = & \vdots \\
0 & + & 0 & + & \cdots & + & f_n F_n & + & (-s_n)\Delta_x & + & (-c_n)\Delta_y & = & 0 \\
c_1 & + & c_2 & + & \cdots & + & c_n & + & 0 & + & 0 & = & 0 \\
s_1 & + & s_2 & + & \cdots & + & s_n & + & 0 & + & 0 & = & W
\end{array}
$$

where

$$c_1 = \cos\theta_1 \qquad c_2 = \cos\theta_2 \qquad c_3 = \cos\theta_3 \qquad c_i = \cos\theta_i$$

and

$$s_1 = \sin\theta_1 \qquad s_2 = \sin\theta_2 \qquad s_3 = \sin\theta_3 \qquad s_i = \sin\theta_i$$

3.1.1 Program to Analyze Multiple Rods With Common Load Point

The program developed to solve the problem discussed in the previous section is summarized in Table 3.1 and demonstrates MATLAB's intrinsic capability to solve simultaneous equations. The output from an example problem consisting of five rods is provided in Section 3.1.3.

Routine	Line	Operation
multirod		script file to execute program.
	27	select one of the example problems.
	28-33	define the input parameters.
	36	zero the coefficient matrix and right hand side vector.
	40	calculate flexibility coefficients.
	43-47	set up the coefficient matrix.
	48	use MATLAB to solve simultaneous equations.
	51-52	perform equilibrium check.
	54-68	output the results.

Table 3.1. Description of Code in Example **multirod**

3.1.2 Exercises

1. Modify the program **multirod** to account for the fact that the angles for θ do not remain constant. That is, modify program **multirod** to incorporate an iterative process similiar to the method employed with example program **truss** in Chapter 1. Assume that the horizontal distances between the ceiling supports (S_1, S_2, S_n) are known and the original location of the load point is known.

3.1.3 Program Output and Code

Output from Example multirod

```
Analysis of Muliple Rods
------------------------

Rod # 1
  Force:        2705.99
  Length:       100
  Flex. Coef.:  3.33333e-05

Rod # 2
  Force:        672.202
  Length:       141.421
  Flex. Coef.:  6.67082e-05
```

3.1. Multiple Rods with a Common Load Point

```
Rod # 3
  Force:        1426.46
  Length:       115.47
  Flex. Coef.:  6.415e-05

Rod # 4
  Force:        5328
  Length:       106.418
  Flex. Coef.:  1.4189e-05

Rod # 5
  Force:        1686.05
  Length:       292.38
  Flex. Coef.:  3.32251e-05

Final Location of Load Point

  Delta x (+ right): -0.0267844
  Delta y (+ down):   0.0901996

Equilibrium Check

  Sum Fx: 0
  Sum Fy: -3.41061e-13
```

Script File multirod

```
 1: % Example: multirod
 2: % ~~~~~~~~~~~~~~~~~
 3: % This example analyzes the effects of a load
 4: % applied to the point of interconnect for
 5: % several rods.  The user defines the
 6: % characteristics of each rod (theta, area, E).
 7: % The resulting displacement of the point
 8: % where the gravity load is applied is
 9: % determined.
10: %
11: % Data is defined in the declaration statements
12: % below, where:
13: %
14: % W      - gravity load
15: % D      - vertical height of rods
```

```
16: % Theta  - vector containing the angle in
17: %          degrees for each rod measured
18: %          from positive y-axis, + is CCW
19: % Area  - vector containing area of each rod
20: % E     - vector containing E for each rod
21: %
22: % User m functions required: none.
23: %----------------------------------------------
24:
25: clear;
26: %...Input definitions
27: Problem=1;
28: if Problem == 1
29:   W=10000; D=100;
30:   Theta=[0     45      -30    20     -70 ];
31:   Area= [0.1   0.2     0.15   0.25   0.4 ];
32:   E=    [30e6  10.6e6  12e6   30e6   22e6];
33: end
34: degrad=pi/180; Theta=Theta*degrad;
35: loop=length(E); loop1=loop+1; loop2=loop+2;
36: sintheta=sin(Theta); costheta=cos(Theta);
37: b=zeros(loop2,1); A=zeros(loop2,loop2);
38:
39: %...Rod length and flexibility coefficient
40: Lrod=D./costheta; Flex=Lrod./(Area.*E);
41:
42: %...Setup and solve simultaneous equations
43: A=diag(Flex);
44: A(1:loop,loop1)=-sintheta';
45: A(1:loop,loop2)=-costheta';
46: A(loop1,1:loop)= costheta;
47: A(loop2,1:loop)= sintheta;
48: b(loop1)=W; x=A\b;
49:
50: %...Equilibrium check
51: sumFx=sum(x(1:loop).*sintheta');
52: sumFy=W-sum(x(1:loop).*costheta');
53:
54: fprintf('\n\nAnalysis of Muliple Rods');
55: fprintf( '\n------------------------');
56: for i=1:loop
57:   fprintf('\n\nRod # %g',i);
58:   fprintf('\n   Force:      %g',x(i));
59:   fprintf('\n   Length:     %g',Lrod(i));
```

3.2. Rigid Bar Suspended by Cables

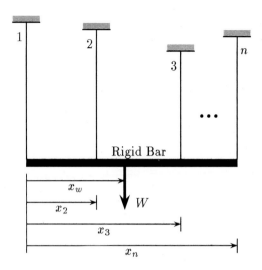

Figure 3.3. Rigid Bar Suspended by Cables

```
60:    fprintf('\n   Flex. Coef.: %g',Flex(i));
61: end
62: fprintf('\n\nFinal Location of Load Point\n');
63: fprintf(   '\n   Delta x (+ right): %g',x(loop1));
64: fprintf(   '\n   Delta y (+ down):  %g',x(loop2));
65: fprintf('\n\nEquilibrium Check\n');
66: fprintf(   '\n   Sum Fx: %g',sumFx);
67: fprintf(   '\n   Sum Fy: %g',sumFy);
68: fprintf('\n');
```

3.2 Rigid Bar Suspended by Cables

A rigid bar is suspended by a set of cables hung from a ceiling as shown in Figure 3.3. The rigid bar is subjected to a gravity load W which is located at the position x_w and the location of W is restricted to ranges which produce either zero or tensile forces in all the cables. The cables have unique properties for area, modulus of elasticity, and length which are given along with the x location of each cable.

The gravity loading will cause the rigid bar to undergo a deflection which is shown in Figure 3.4. The deflection of each cable can be determined if the force in the cable is known. However, there are only two equations from statics available and there may be several cables. The problem is statically indeterminate and additional equations must be developed relating the deformation characteristics of the system.

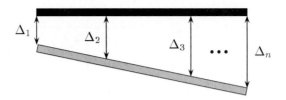

Figure 3.4. Deflection of Rigid Bar

There are two equations which can be written using statics, or

$$\sum F_y = 0, \qquad \sum_{i=1}^{n} F_i = W$$

$$\sum M_1 = 0, \qquad \sum_{i=2}^{n} F_i x_i = W x_w$$

The relationship between the deflections of the rods can be written using Figure 3.4 and by recognizing that the rigid bar remains straight. The deflections are linearly related by

$$\Delta_i = \Delta_1 + \frac{x_i}{x_n}(\Delta_n - \Delta_1) \qquad i = 1, \ldots, n$$

This equation implies that the deflection of the left and right cables is sufficient to determine the deflection for each of the other cables. The deflection of each axially loaded cable can be determined using

$$\Delta_i = \left(\frac{FL}{AE}\right)_i \qquad i = 1, \ldots, n$$

where A is the cable area, L is the cable length, and E is the modulus of elasticity for a cable. Rearranging to calculate the force in terms of the deflection yields

$$F_i = \left(\frac{AE}{L}\right)_i \Delta_i \qquad \text{or} \qquad F_i = k_i \Delta_i \qquad i = 1, \ldots, n$$

where k is called the *stiffness coefficient*. Substituting the expression for deflection into this equation and rearranging terms produces

$$F_i = k_i \left[\left(1 - \frac{x_i}{x_n}\right)\Delta_1 + \left(\frac{x_i}{x_n}\right)\Delta_n\right] \qquad i = 1, \ldots, n$$

This equation relates the unknown force in each cable with the deflections at only two cables (the left and right cables). Substituting this expression for F_i into the

3.2. Rigid Bar Suspended by Cables

two equations from statics yields

$$\overbrace{\left[\sum_{i=1}^{n} k_i \left(1 - \frac{x_i}{x_n}\right)\right]}^{a_{11}} \Delta_1 + \overbrace{\left[\sum_{i=1}^{n} k_i \left(\frac{x_i}{x_n}\right)\right]}^{a_{12}} \Delta_n = W$$

$$\underbrace{\left[\sum_{i=1}^{n} x_i k_i \left(1 - \frac{x_i}{x_n}\right)\right]}_{a_{21}} \Delta_1 + \underbrace{\left[\sum_{i=1}^{n} x_i k_i \left(\frac{x_i}{x_n}\right)\right]}_{a_{22}} \Delta_n = W x_w$$

This provides two equations with two unknowns, Δ_1 and Δ_n. Once these two deflections are determined the deflections and resulting forces can be calculated for the other cables.

Recall that the gravity load is restricted to a range of locations which only induce zero or tensile forces in the cables. This range of values can easily be identified. First, set the deflection for the left cable, Δ_1, to zero. The first equation of the two simultaneous equations can be immediately solved for the deflection at the right cable, Δ_n. This value (and $\Delta_1 = 0$) may be substituted into the second equation to determine the extreme right position of the gravity load (that is, solve for x_w). A similar process is followed to determine the extreme left position. If this precedure is performed algebraically the following simple equations result.

$$x_\text{left} = \frac{a_{21}}{a_{11}} \qquad x_\text{right} = \frac{a_{22}}{a_{12}}$$

3.2.1 Program to Analyze a Rigid Bar Suspended by Cables

A program was developed to solve the problem discussed in the previous section and is summarized in Table 3.2. The output from an example problem consisting of four cables is provided in Section 3.2.3. Figure 3.5 shows the position of the rigid bar which results from the gravity load being placed at several positions within the allowable range. Note that the plot indicates there is a point along the rigid bar which deflects the same amount regardless of where the load is placed (at approximately $x = 72$ for the example problem). Figure 3.6 contains a three-dimensional plot of the same information. However, this plot includes the additional axis showing the location of the load. The reader should note that this plot creates the appearance that the rigid bar does not remain straight. This results from the fact that the aspect ratios for the three axes are not consistent.

Two MATLAB programs are included in Section 3.2.3 to solve this problem. The first program, **rbar1**, demonstrates the ability to perform operations directly on vectors and matrices using MATLAB's intrinsic vector and matrix operators. The use of these operators greatly improves the clarity of the program, reduces the number of program statements, and allows MATLAB to utilize the most efficient (and thusly fastest) internal code to perform these operations. The use of these operators is commonly referred to as *vectorization* of the code. The second program,

rbar2, is an analogous set of code which does not employ the vector operators. The reader should carefully study these two programs side-by-side to understand the use of vector and matrix operators.

Routine	Line	Operation
rbar1		script file to execute program.
	27	select one of the example problems.
	28-34	define the input parameters.
	37-39	calculate some values. Notice the use of the element-by-element vector operator.
	41-46	set up the coefficients for the simultaneous equations.
	48-49	calculate the extreme positions for the load.
	56-57	define the positions where the load will be placed.
	58-65	solve for the deflections, forces, and stresses resulting from each location of the load.
	60	solve the simultaneous equations using MATLAB operator. Notice the use of the transpose operator.
	68-82	output the results.
	84-90	create the 2D plot of rod positions.
	92-99	create the 3D plot of rod positions.
	98	change the viewing angle for the plot.

Table 3.2. Description of Code in Example **rbar1**

3.2. Rigid Bar Suspended by Cables

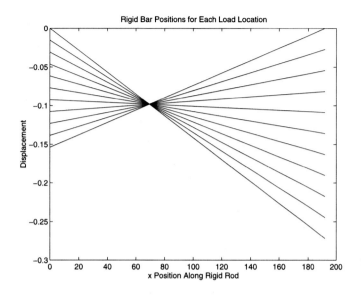

Figure 3.5. Rigid Bar Positions for Each Load Location in xy

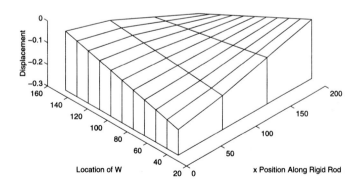

Figure 3.6. Rigid Bar Positions for Each Load Location in xyz

3.2.2 Exercises

1. Add the capability to program **rbar1** to calculate the location which the load should be placed to cause the deflection of all the cables to be equal. This will cause the rigid bar to remain horizontal. (Hint: Study how the two points determining the allowable range of loads was found.)

2. Modify the program **rbar1** by placing the code for determining the values of displacement, force, and stress within a separate function. Then use these functions to develop a program which will determine these values for a user specified set of loading positions.

3. Figure 3.5 indicates that for a given problem there is a particular point along the bar which maintains the same displacement regardless of where the load is placed (or, the rod pivots about this point). Add the capability to program **rbar1** to determine this point automatically.

3.2.3 Program Output and Code

Output from Example rbar1

```
Rigid Bar Suspended by Cables
-----------------------------

Acceptable Range of Load Point

  X (left):   28.3454
  X (right):  141.94

P at x = 28.3454
  Cable # 1
    Force:        64140.7
    Stress:       64140.7
    Deflection:  -0.153938
  Cable # 2
    Force:        20396.8
    Stress:       10198.4
    Deflection:  -0.115453
  Cable # 3
    Force:        15462.5
    Stress:       20616.7
    Deflection:  -0.0577267
  Cable # 4
    Force:        0
    Stress:       0
    Deflection:   0
```

3.2. Rigid Bar Suspended by Cables

```
P at x = 39.7049
  Cable # 1
    Force:        57726.7
    Stress:       57726.7
    Deflection: -0.138544
  Cable # 2
    Force:        19557.8
    Stress:       9778.91
    Deflection: -0.110705
  Cable # 3
    Force:        18467.6
    Stress:       24623.5
    Deflection: -0.0689457
  Cable # 4
    Force:        4247.92
    Stress:       8495.83
    Deflection: -0.0271867


  .
  .  Material removed
  .

P at x = 141.94
  Cable # 1
    Force:        0
    Stress:       0
    Deflection: 0
  Cable # 2
    Force:        12007.4
    Stress:       6003.72
    Deflection: -0.0679667
  Cable # 3
    Force:        45513.4
    Stress:       60684.5
    Deflection: -0.169917
  Cable # 4
    Force:        42479.2
    Stress:       84958.3
    Deflection: -0.271867
```

Script File rbar1

```
1:  % Example: rbar1
2:  % ~~~~~~~~~~~~~~
3:  % This example calculates the deflection of
4:  % a rigid bar suspended by a set of cables.
5:  % This version uses the vector and matrix
6:  % operators of MATLAB.
7:  %
8:  % Data is defined in the declaration statements
9:  % below, where:
10: %
11: % W       - gravity load
12: % Xcable  - vector containing x location for
13: %           each cable
14: % Lcable  - vector containing cable lengths
15: % Acable  - vector containing cable areas
16: % Ecable  - vector containing cable Es
17: % No_W_load_pts - number of solutions to
18: %           generate for different positions
19: %           of applied W
20: %
21: % User m functions required:
22: %     genprint
23: %-----------------------------------------------
24:
25: clear;
26: %...Input definitions
27: Problem=1;
28: if Problem == 1
29:    W=100000; No_W_load_pts=10;
30:    Xcable=[   0       48     120    192 ];
31:    Lcable=[  72      120      84     96 ];
32:    Acable=[   1        2    0.75    0.5 ];
33:    Ecable=[30e6    10.6e6   30e6   30e6];
34: end
35: No_cables=length(Ecable);
36:
37: %...Calculate AE/L and X ratio for each cable
38: Cael=(Acable.*Ecable)./Lcable;
39: Xratio=Xcable./Xcable(No_cables);
40:
41: %...Setup and solve simultaneous equations
```

3.2. Rigid Bar Suspended by Cables

```
42: A(1,1)=sum(Cael-Xratio.*Cael);
43: A(1,2)=sum(Xratio.*Cael);
44: A(2,1)=sum((Xcable.*Cael).*(-Xratio+1));
45: A(2,2)=sum(Xcable.*Xratio.*Cael);
46: b(1)=W;
47:
48: %...Solve for extreme locations of W
49: Xleft=A(2,1)/A(1,1); Xright=A(2,2)/A(1,2);
50:
51: %..........................................
52: %...Loop to solve for deflection, force, and
53: %...stress in each cable as load is moved
54: %...across allowable range
55: %..........................................
56: X_inc=(Xright-Xleft)/No_W_load_pts;
57: Xw=Xleft:X_inc:Xright;
58: for j=1:No_W_load_pts+1
59:    %...Solve simultaneous equations
60:    b(2)=W*Xw(j); x=A\b';
61:    %...Calculate remaining quantities
62:    Delta(j,:)=(Xratio*(x(2)-x(1)))+x(1);
63:    Fcable(j,:)=Cael.*Delta(j,:);
64:    Stress(j,:)=Fcable(j,:)./Acable;
65: end
66: Delta=-Delta;
67:
68: fprintf('\nRigid Bar Suspended by Cables');
69: fprintf('\n-----------------------------');
70: fprintf('\n\nAcceptable Range of Load Point\n');
71: fprintf(  '\n   X (left):  %g',Xleft);
72: fprintf(  '\n   X (right): %g',Xright);
73: for j=1:No_W_load_pts+1
74:    fprintf('\n\nW at x = %g',Xw(j));
75:    for i=1:No_cables
76:       fprintf('\n   Cable # %g',i);
77:       fprintf('\n      Force:      %g',Fcable(j,i));
78:       fprintf('\n      Stress:     %g',Stress(j,i));
79:       fprintf('\n      Deflection: %g',Delta(j,i));
80:    end
81: end
82: fprintf('\n');
83:
84: clf; plot(Xcable,Delta);
85:    title(['Rigid Bar Positions for Each' ...
```

```
86:              ' Load Location']);
87:     xlabel('x Position Along Rigid Rod');
88:     ylabel('Displacement'); drawnow;
89: % genprint('rod-xy');
90:     disp('Press key to continue'); pause
91:
92: clf; meshz(Xcable,Xw,Delta);
93:     title(['Rigid Bar Positions for Each' ...
94:              ' Load Location']);
95:     xlabel('x Position Along Rigid Rod');
96:     ylabel('Location of W');
97:     zlabel('Displacement');
98:     view(-45,60); drawnow;
99: % genprint('mesh-rod');
```

Script File rbar2

```
 1: % Example: rbar2
 2: % ~~~~~~~~~~~~~~
 3: % This example calculates the deflection of
 4: % a rigid bar suspended by a set of cables.
 5: % This version uses the regular loop
 6: % indexing.
 7: %
 8: % Data is defined in the declaration statements
 9: % below, where:
10: %
11: % W       - gravity load
12: % Xcable - vector containing x location for
13: %          each cable
14: % Lcable - vector containing cable lengths
15: % Acable - vector containing cable areas
16: % Ecable - vector containing cable Es
17: % No_W_load_pts - number of solutions to
18: %          generate for different positions
19: %          of applied W
20: %
21: % User m functions required:
22: %     genprint
23: %-----------------------------------------
24:
25: clear;
```

3.2. Rigid Bar Suspended by Cables

```
26: %...Input definitions
27: Problem=1;
28: if Problem == 1
29:   W=100000; No_W_load_pts=10;
30:   Xcable=[    0        48      120     192 ];
31:   Lcable=[   72       120       84      96 ];
32:   Acable=[    1         2     0.75     0.5 ];
33:   Ecable=[30e6     10.6e6     30e6    30e6];
34: end
35: No_cables=length(Ecable);
36:
37: %...Calculate AE/L and X ratio for each cable
38: for i=1:No_cables
39:   Cael(i)=(Acable(i)*Ecable(i))/Lcable(i);
40:   Xratio(i)=Xcable(i)/Xcable(No_cables);
41: end
42:
43: %...Setup and solve simultaneous equations
44: A=zeros(2,2);
45: for i=1:No_cables
46:   temp=Xratio(i)*Cael(i);
47:   A(1,1)=A(1,1)+Cael(i)-temp;
48:   A(1,2)=A(1,2)+temp;
49:   A(2,1)=A(2,1)+Xcable(i)*Cael(i)*(1-Xratio(i));
50:   A(2,2)=A(2,2)+Xcable(i)*temp;
51: end
52: b(1)=W;
53:
54: %...Solve for extreme locations of W
55: Xleft=A(2,1)/A(1,1); Xright=A(2,2)/A(1,2);
56:
57: %........................................
58: %...Loop to solve for deflection, force, and
59: %...stress in each cable as load is moved
60: %...across allowable range
61: %........................................
62: X_inc=(Xright-Xleft)/No_W_load_pts;
63: Xw=Xleft:X_inc:Xright;
64: for j=1:No_W_load_pts+1
65:   %...Solve simultaneous equations
66:   b(2)=W*Xw(j); x=A\b';
67:   %...Calculate remaining quantities
68:   for i=1:No_cables
69:     Delta(j,i)=(1-Xratio(i))*x(1)+ ...
```

```
70:              Xratio(i)*x(2);
71:       Fcable(j,i)=Cael(i)*Delta(j,i);
72:       Stress(j,i)=Fcable(j,i)/Acable(i);
73:    end
74: end
75: Delta=-Delta;
76:
77: fprintf('\nRigid Bar Suspended by Cables');
78: fprintf('\n----------------------------');
79: fprintf('\n\nAcceptable Range of Load Point\n');
80: fprintf(  '\n   X (left):  %g',Xleft);
81: fprintf(  '\n   X (right): %g',Xright);
82: for j=1:No_W_load_pts+1
83:    fprintf('\n\nW at x = %g',Xw(j));
84:    for i=1:No_cables
85:       fprintf('\n   Cable # %g',i);
86:       fprintf('\n      Force:      %g',Fcable(j,i));
87:       fprintf('\n      Stress:     %g',Stress(j,i));
88:       fprintf('\n      Deflection: %g',Delta(j,i));
89:    end
90: end
91: fprintf('\n');
92:
93: clf; plot(Xcable,Delta);
94:    title(['Rigid Bar Positions for Each' ...
95:           ' Load Location']);
96:    xlabel('x Position Along Rigid Rod');
97:    ylabel('Displacement'); drawnow;
98: % genprint('rod-xy');
99:    disp('Press key to continue'); pause
100:
101: clf; meshz(Xcable,Xw,Delta);
102:    title(['Rigid Bar Positions for Each' ...
103:           ' Load Location']);
104:    xlabel('x Position Along Rigid Rod');
105:    ylabel('Location of W');
106:    zlabel('Displacement');
107:    view(-45,60); drawnow;
108: % genprint('mesh-rod');
```

3.3. Axially Loaded Shaft

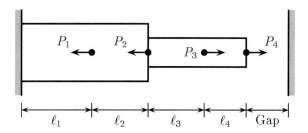

Figure 3.7. Axially Loaded Shaft

3.3 Axially Loaded Shaft

The previous two statically indeterminate problems required developing a set of equations which were solved simultaneously to determine the unknowns. In this section the unit load method will be employed to analyze a statically indeterminate axially loaded shaft. Figure 3.7 depicts an axial shaft constructed of several segments each having unique properties for length, area, applied axial load, modulus of elasticity, and coefficient of thermal expansion. The shaft is attached to a fixed support on the left side. A gap exists between the right extreme of the shaft and the right support. If the gap is closed due to the loading the problem becomes statically indeterminate.

The unit load method is one method which can be used to analyze this problem. First, the deflection at the right of the shaft is determined by assuming the right wall does not exist (a static analysis). If the deflection is less than the amount specified as the gap then the solution is complete. If the deflection is greater than the value for the gap then additional calculations must be performed. These calculations are aimed at determining the amount of force which must be applied to the right end of the shaft to *push* it back to the point of the right wall. In other words, this is the force which the wall would provide to prevent the shaft from going through the wall. Once the force from the right wall has been determined the system becomes statically determinate.

The analysis begins by assuming that the right wall does not exist. Therefore, the sum of the applied forces yields the reaction at the left wall, or

$$P_{\text{total}} = \sum_{i=1}^{n} P_i$$

where n is the number of unique segments in the shaft and P_i is the applied force at the right end of each segment. The force in each segment of the shaft can be determined by constructing a free-body diagram of each segment, or

$$\tilde{F}_1 = P_{\text{total}}$$

$$\tilde{F}_i = \tilde{F}_{i-1} - P_{i-1} \qquad i = 2, \ldots, n$$

where $\tilde{F}_i$ represents the internal force in each segment for the statically determinate system. The deflection due to the change in temperature can be calculated as

$$(\Delta_\alpha)_i = \alpha_i (\Delta T)_i L_i \qquad i = 1, \ldots, n$$

where α is the coefficient of thermal expansion, L is the length, and ΔT is the change in temperature. Next, the deflection for each segment in the shaft is calculated for a *unit load* applied at the right end of the shaft, or

$$(\Delta_u)_i = \frac{L_i}{A_i E_i} \qquad i = 1, \ldots, n$$

where A is the area and E is the modulus of elasticity. The deflection due to the internal forces in the statically determinate system are now calculated using

$$(\Delta_f)_i = \tilde{F}_i (\Delta_u)_i \qquad i = 1, \ldots, n$$

Summing the contributions for each of these deflection equations yields

$$(\Delta_\alpha)_{\text{total}} = \sum_{i=1}^{n} (\Delta_\alpha)_i$$

$$(\Delta_u)_{\text{total}} = \sum_{i=1}^{n} (\Delta_u)_i$$

$$(\Delta_f)_{\text{total}} = \sum_{i=1}^{n} (\Delta_f)_i$$

Therefore, the deflection of the statically determinate system is

$$\tilde{\Delta}_{\text{total}} = (\Delta_\alpha)_{\text{total}} + (\Delta_f)_{\text{total}}$$

The question which must be answered is whether the deflection resulting from the analysis of the statically determinate system is sufficient to close the gap[1]. If the gap does not close the solution is complete and the reactions at the left and right walls are

$$R_{\text{left}} = -P_{\text{total}} \qquad R_{\text{right}} = 0$$

If the gap has closed then the system is indeterminate and the force necessary to *push* the shaft back to the location of the right wall can be calculated by

$$R_{\text{right}} = -\frac{\tilde{\Delta}_{\text{total}} - \text{GAP}}{(\Delta_u)_{\text{total}}}$$

---

[1] The reader should note that the procedure outlined can be used to solve statically determinate configurations. If the value for the gap is set sufficiently large to insure that the gap never closes then the procedure produces the solution for the statically determinate case.

3.3. Axially Loaded Shaft

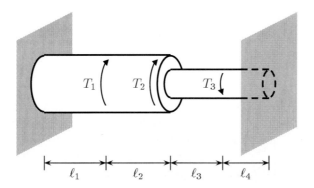

Figure 3.8. Torsionally Loaded Shaft

and the reaction at the left wall is determined from statics.

$$R_{\text{left}} = -(P_{\text{total}} + R_{\text{right}})$$

Once the reactions have been calculated for the indeterminate system the internal force in each segment of the shaft may be found using

$$F_1 = -R_{\text{left}}$$

$$F_i = F_{i-1} - P_{i-1} \qquad i = 2, \ldots, n$$

3.3.1 Program to Analyze an Axially Loaded Shaft

The procedure just described was implemented in program **indaxial** and a summary of the code is contained in Table 3.3. The output from an example problem is provided in Section 3.3.3. The program produces a plot of the original geometry which is not included in the text.

3.3.2 Exercises

1. Figure 3.8 depicts a torsionally loaded shaft. The analysis of this problem is similar to the analysis of an axially loaded shaft. Write a program to analyze a torsionally loaded shaft. The program should allow the user to specify an initial "slop" angle. In other words, the shaft can twist in either direction by this amount prior to catching against a stop mounted on the right wall.

Routine	Line	Operation
indaxial		script file to execute program.
	28	select one of the example problems.
	29-53	define the input parameters.
	57	use intrinsic function **sum** to total the applied load.
	60-63	calculate the force in each segment assuming the right wall does not exist.
	67-69	calculate the total deflection due to temperature, axial load, and unit axial load.
	70	Compute the total deflection.
	74-75	compute the reactions if the gap did not close.
	77-78	compute the reactions if the gap did close.
	82-85	compute the actual force in each segment.
	86	use the element-by-element vector operator to calculate the stress in each segment.
	88-114	output the results.
	123-137	create a plot of the geometry.

Table 3.3. Description of Code in Example **indaxial**

3.3.3 Program Output and Code

Output from Example indaxial

```
            Analysis of
Axially Loaded Indeterminate Member
-----------------------------------

Gap: 0.02  (*** Gap Closed ***)

Section: 1
   Area:         2.5
   Length:       12
   Alpha:        1.01e-05
   E:            1.5e+07
   Delta t:      200
   Applied P:    0
   Internal P:   -52000
   Stress:       -20800
```

3.3. Axially Loaded Shaft

```
Section: 2
  Area:        3
  Length:      15
  Alpha:       1.28e-05
  E:           1e+07
  Delta t:     200
  Applied P:   0
  Internal P:  -52000
  Stress:      -17333.3

Reaction (left wall): 52000
Reaction (right wall): -52000
```

Script File indaxial

```
 1: % Example: indaxial
 2: % ~~~~~~~~~~~~~~~~~
 3: % This program calculates the internal forces
 4: % and stresses in an axially loaded
 5: % indeterminate member.
 6: %
 7: % Data is defined in the declaration statements
 8: % below, where:
 9: %
10: % Gap       - initial gap
11: % Area      - vector with each segment's area
12: % Length    - vector with each segment's length
13: % Alpha     - vector with each segment's
14: %             coefficient of thermal expansion
15: % Emod      - vector with each segment's E
16: % Temp_diff - vector with temperature
17: %             differential for each segment
18: % Applied_P - vector of axial loads (always
19: %             applied on the left end of a
20: %             segment
21: %
22: % User m functions required:
23: %     genprint
24: %-------------------------------------------------
25:
26: clear;
27: %...Input definitions
28: Problem=1;
```

```matlab
29: if Problem == 1
30:   %...Thermal load only
31:   Gap=0.02;
32:   Area=     [2.5      3      ];
33:   Length=   [12       15     ];
34:   Alpha=    [10.1e-6  12.8e-6];
35:   Emod=     [15e6     10e6   ];
36:   Temp_diff=[200      200    ];
37:   Applied_P=[0        0      ];
38: elseif Problem == 2
39:   Gap=0.1;
40:   Area=[3 3 2 2 4];
41:   Length=[20 20 30 30 40];
42:   Alpha=[10.1e-6 10.1e-6 ...
43:          12.8e-6 12.8e-6 6.5e-6];
44:   Emod=[15e6 15e6 10e6 10e6 30e6];
45:   Temp_diff=[100 100 75 75 110];
46:   Applied_P=[2e4 0 -5e3 3e4 0];
47: elseif Problem == 3
48:   %...Thermal load only
49:   Gap=0;
50:   Area=     [0.75   ]; Length=  [30    ];
51:   Alpha=    [6.5e-6 ]; Emod=    [30e6  ];
52:   Temp_diff=[-200   ]; Applied_P=[0    ];
53: end
54: No_segs=length(Area); Gap_Flag=0;
55:
56: %...Total applied load
57: Sum_P=sum(Applied_P);
58:
59: %...Force in each section if right wall removed
60: Fseg(1)=Sum_P;
61: for i=2:No_segs
62:   Fseg(i)=Fseg(i-1)-Applied_P(i-1);
63: end
64:
65: %...Deflection due to temperature, PL/AE, and
66: %...unit load
67: Delta_unit_total=sum(Length./(Area.*Emod));
68: Delta_plae_total=sum(Fseg.*Length./(Area.*Emod));
69: Delta_total=sum(Alpha.*Temp_diff.*Length)+ ...
70:             Delta_plae_total;
71:
72: %...Has the gap closed, find wall reactions
```

3.3. Axially Loaded Shaft

```
73: if (Gap ~= 0) & (Delta_total < Gap)
74:   %...Gap did not close
75:   Gap_Flag=1; Rleft=-Sum_P; Rright=0;
76: else
77:   Rright=-(Delta_total-Gap)/Delta_unit_total;
78:   Rleft=-(Sum_P+Rright);
79: end
80:
81: %...Find true force and stresses
82: True_F(1)=-Rleft;
83: for i=2:No_segs
84:   True_F(i)=True_F(i-1)-Applied_P(i-1);
85: end
86: Stress=True_F./Area;
87:
88: fprintf('\n\n              Analysis of');
89: fprintf( ...
90:   '\nAxially Loaded Indeterminate Member');
91: fprintf( ...
92:   '\n--------------------------------');
93: fprintf('\n\nGap: %g',Gap);
94: if Gap_Flag == 1
95:   fprintf('  (*** Gap did not close ***)');
96:   fprintf('\n    Total deflection: %g', ...
97:     Delta_total);
98: else
99:   fprintf('  (*** Gap closed ***)');
100: end;
101: for i=1:No_segs
102:   fprintf('\n\nSection: %g',i);
103:   fprintf( '\n   Area:        %g',Area(i));
104:   fprintf( '\n   Length:      %g',Length(i));
105:   fprintf( '\n   Alpha:       %g',Alpha(i));
106:   fprintf( '\n   E:           %g',Emod(i));
107:   fprintf( '\n   Delta t:     %g',Temp_diff(i));
108:   fprintf( '\n   Applied P:   %g',Applied_P(i));
109:   fprintf( '\n   Internal P:  %g',True_F(i));
110:   fprintf( '\n   Stress:      %g',Stress(i));
111: end
112: fprintf('\n\nReaction (left wall): %g',Rleft);
113: fprintf('\nReaction (right wall): %g',Rright);
114: fprintf('\n');
115:
116: %...Plot the geometry
```

```
117: %....................
118: %...Force axes to contain entire plot since
119: %...hold is being used
120: x(1)=0; x(2)=sum(Length)+Gap;
121: y(1)=-0.5; y(2)=0.5; Amax=max(Area);
122:
123: clf;
124: plot(x,y,'.'); hold on;
125:   xlabel('x'); ylabel('Area ratios');
126:   title('Indeterminate Member Geometry');
127:   for i=1:No_segs
128:     Left_x=sum(Length(1:i))-Length(i);
129:     Right_x=sum(Length(1:i));
130:     xp=[Left_x Right_x Right_x Left_x Left_x];
131:     Bot_y=-(Area(i)/Amax)/2;
132:     Top_y= (Area(i)/Amax)/2;
133:     yp=[Bot_y Bot_y Top_y Top_y Bot_y];
134:     plot(xp,yp,'-');
135:   end
136:   hold off; drawnow;
137: % genprint('axialgeo');
```

Chapter 4

Geometrical Properties of Polygons

Many problems encountered in mechanics require the accurate calculation for the geometrical properties of a cross section. Typically, the practitioner uses one of the many references [13, 38] which contain tables of properties for dozens of individual cross sections. However, even comprehensive reference books contain only a few dozen cross sections. Wilson and Farrior [51] published one of the first papers on the automatic calculation of geometrical properties for cross sections which can be described using polygons. Subsequently, many papers have been published concerning this important topic [4, 12, 20, 22, 27, 50, 52]. In this chapter a computer program to determine geometric properties for polygons is presented. Additionally, an instructional program which determines the approximate value for the polar moment of inertia for a circular cross section is presented. A polygon is used as an approximation for the circular cross section. This example demonstrates the accuracy of replacing a curved boundary by an n-sided polygon.

4.1 Properties of General Areas

For a general area, as shown in Figure 4.1, the fundamental definitions for the geometrical properties of a cross section are the area,

$$A = \int_A dA \qquad \text{where} \qquad dA = dx\ dy$$

the first moments of the area (Q_x and Q_y) and centroids ($\bar{x}$ and $\bar{y}$),

$$Q_x = \int_A y\ dA \qquad \bar{y} = \frac{Q_x}{A}$$

$$Q_y = \int_A x\ dA \qquad \bar{x} = \frac{Q_y}{A}$$

and the second moments of the area, also called the rectangular moments of inertia.

$$I_x = \int_A y^2\ dA$$

$$I_y = \int_A x^2 \, dA$$

$$I_{xy} = \int_A xy \, dA$$

Inspection of these relationships indicates that values for Q_x and Q_y may be positive, negative, or zero. Additionally, I_x and I_y will always be positive while I_{xy}, the product of inertia, may be positive, negative, or zero. If the value for I_{xy} is zero then one or both of the centroidal axes are an axis of symmetry.

Another important geometric property is the polar moment of inertia which is always a positive value and can be defined as

$$\begin{aligned} J &= \int_A \rho^2 \, dA \\ &= \int_A (x^2 + y^2) \, dA \\ &= I_y + I_x \end{aligned}$$

A final set of geometric properties, the *radii of gyration*[1], are defined as

$$r_x = \left(\frac{I_x}{A}\right)^{1/2} \qquad r_y = \left(\frac{I_y}{A}\right)^{1/2}$$

$$r_J = \left(\frac{J}{A}\right)^{1/2} \qquad r_J^2 = r_x^2 + r_y^2$$

where r_x and r_y are the rectangular radii of gyration and r_J is the polar radius of gyration.

4.1.1 Parallel-Axis Theorem

The inertial moments of an area depend on the reference axes chosen. Using the moments of inertia about centroidal axes, the *parallel-axis theorem* (or transfer formula), may be employed to find the moments about any parallel axis as shown in Figure 4.2, or

$$I'_x = \bar{I}_x + Ay^2$$

$$I'_y = \bar{I}_y + Ax^2$$

$$I'_{xy} = \bar{I}_{xy} + Axy$$

$$J' = \bar{J} + A\rho^2$$

---

[1] The radius of gyration is analogous to the location at which the entire area could be concentrated and still produce the same moment of inertia as the original cross section.

4.1. Properties of General Areas

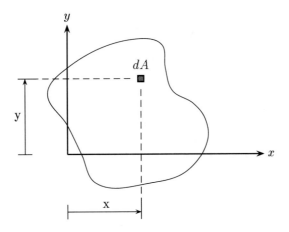

Figure 4.1. First and Second Moment of an Area

where the quantities $\bar{I}_x$, $\bar{I}_y$, $\bar{J}$, and $\bar{I}_{xy}$ represent the set of inertial values for the centroidal axes. These equations lead to the following relationships for determining the inertial values for the centroidal set of axes when the geometrical properties are known about a non-centroidal set of axes.

$$\bar{I}_x = I'_x - A(\bar{y})^2$$

$$\bar{I}_y = I'_y - A(\bar{x})^2$$

$$\bar{I}_{xy} = I'_{xy} - A\bar{x}\bar{y}$$

$$\bar{J} = J - A(\bar{\rho})^2$$

The set of relations for the radii of gyration can also be transferred using the parallel-axis theorem, or

$$(r'_x)^2 = (\bar{r}_x)^2 + y^2$$

$$(r'_y)^2 = (\bar{r}_y)^2 + x^2$$

$$(r'_J)^2 = (\bar{r}_J)^2 + \rho^2$$

It is important to realize that the parallel-axis theorem only applies when one of the two coordinate reference systems is the centroidal system. To shift properties between arbitrary parallel axes, axes should first be shifted from the initial axes to

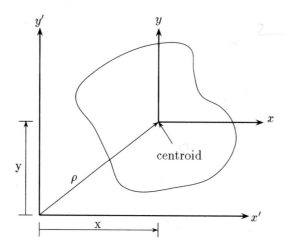

Figure 4.2. Parallel-Axis Theorem

the centroidal axes. Then a second step can be taken to shift from the centroidal axes to the final position.

4.1.2 Rotation of Inertia

The inertial values are dependent on the angular orientation of the set of axes chosen. Therefore, their values will change as the perpendicular reference axes are rotated about the centroid as shown in Figure 4.3. The relationships for the inertial values when rotated through an angle θ (θ is measured from the positive x-axis and is positive in the counterclockwise direction) are:

$$\hat{I}_x = \frac{I_x + I_y}{2} + \frac{I_x - I_y}{2}\cos(2\theta) - I_{xy}\sin(2\theta)$$

$$\hat{I}_y = \frac{I_x + I_y}{2} - \frac{I_x - I_y}{2}\cos(2\theta) + I_{xy}\sin(2\theta)$$

$$\hat{I}_{xy} = \frac{I_x - I_y}{2}\sin(2\theta) + I_{xy}\cos(2\theta)$$

These equations are called the *transformation equations for moments and products of inertia*. The values of θ which make $\hat{I}_x$ have the largest or smallest values can

4.1. Properties of General Areas

be computed by setting $\frac{d\hat{I}_x}{d\theta}$ equal to zero and solving for the principal angular orientation, θ_p. This produces

$$(I_x - I_y)\sin(2\theta_p) + 2I_{xy}\cos(2\theta_p) = 0$$

Taking

$$\theta_p = \frac{1}{2}\tan^{-1}\left(\frac{-2I_{xy}}{I_x - I_y}\right)$$

gives two values of θ_p which are 90° apart and the angle θ_p is measured from the positive x-axis to the principal planes. These two planes are called the *principal axes of inertia*. The reader should note that the sign for the numerator has been maintained as part of the numerator rather than negating the entire equation (as many texts present this equation). Writing the equation in this fashion allows the value of θ_p to be associated with the proper principal inertial value[2]. Also note that $I_{xy} \neq 0$ and $I_x = I_y$ implies $\theta_p = \pm\pi/4$. One of these angles maximizes I_x and the other minimizes I_x. Additionally, the critical values of θ_p make I_{xy} zero.

It is not hard to verify that the set of all pairs of $(\hat{I}_x, \hat{I}_y)$ lie on a circle defined by

$$\left(\hat{I}_x - \frac{I_x + I_y}{2}\right)^2 + \hat{I}_{xy}^2 = \left(\frac{I_x + I_y}{2}\right)^2 + I_{xy}^2$$

This circle has a radius of

$$R = \sqrt{\left(\frac{I_x + I_y}{2}\right)^2 + I_{xy}^2}$$

and center coordinates $(\frac{I_x+I_y}{2}, 0)$. Therefore the critical values of inertial moment are

$$I_{\max} = \frac{I_x + I_y}{2} + R$$

$$I_{\min} = \frac{I_x + I_y}{2} - R$$

which correspond to $\hat{I}_{xy} = 0$.

The maximum and minimum value for I_{xy} is produced on a plane 45° from the planes for $I_{\max}$ and $I_{\min}$, or

$$(I_{xy})_{\max/\min} = \pm R$$

and I_x and I_y on these planes is non-zero. The equation for the angle to the plane of maximum and minimum I_{xy} is

$$\alpha = \frac{1}{2}\tan^{-1}\left(\frac{I_x - I_y}{2I_{xy}}\right)$$

---

[2] The quadrant in which this angle lies can be determined by the signs of the numerator and denominator.

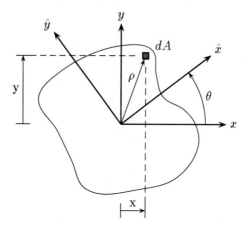

Figure 4.3. Rotation of Inertia About Centroidal Axes

However, in practice this equation is rarely used since the $(I_{xy})_{\max}$ plane is $+45°$ from the plane of $I_{\max}$ and the plane to $(I_{xy})_{\min}$ plane is $-45°$ from the plane of $I_{\max}$. Finally, if the denominator of the above equation is zero (i.e. $I_{xy} = 0$) then the axes are principal inertial axes.

One further comment regarding principal axes should be made. A set of axes are principal axes if and only if I_{xy} is zero for that set. However, a sufficient condition to make I_{xy} zero is that either the x-axis or the y-axis be an axis of symmetry. This circumstance often occurs in practice. Consider, for example, a set of centroidal axes for a rectangle. If the axes are parallel to the rectangle sides, then $I_{xy} = 0$. Hence, I_x and I_y are principal moments of inertia. Furthermore, when the rectangle is a square, the minimum and maximum moments are the same, thereby making the inertial moments the same in all directions. It is evident from the basic equations that $I_x = I_y$ and $I_{xy} = 0$ implies $\hat{I}_x = \hat{I}_y$ and $\hat{I}_{xy} = 0$ for all axes.

4.2 Geometrical Properties of Polygons

The properties A, $\bar{x}$, $\bar{y}$, Q_x, Q_y, I_x, I_y, and I_{xy} for an arbitrary polygon can be developed by using properties of a triangle. It will be shown below that a polygon can be thought of as the superposition of a number of triangles.

First, the properties of an arbitrary triangle having corners at R_1, R_2, R_3 in three-dimensional space are presented. Use of the vector cross product implies

$$\hat{\eta} A = \frac{1}{2}(R_2 - R_1) \times (R_3 - R_1)$$

4.2. Geometrical Properties of Polygons

$$= \frac{1}{2}[R_1 \times R_2 + R_2 \times R_3 + R_3 \times R_1]$$

where A is the triangle area and $\hat{\eta}$ denotes the unit normal to a plane through the tips of R_1, R_2, and R_3. Furthermore, it is known that the centroid of the triangle is located at the average of the corner radii. Thus the centroidal radius is

$$R_c = \frac{R_1 + R_2 + R_3}{3}$$

Finding the inertial moments of the triangle is more involved than obtaining the area and centroid properties just stated. The inertial moment can be defined by the following symmetric matrix

$$\text{ARR} = \int\int RR^T \, d(\text{area})$$

where R is the cartesian radius vector represented as a column. For a triangle this computation yields

$$\text{ARR} = \frac{A}{12}\left(R_1 R_1^T + R_2 R_2^T + R_3 R_3^T + 9 R_c R_c^T\right)$$

and is valid for either two or three dimensions.

The previous relations will be applied to a triangle lying in the xy plane and having corners at (x_1, y_1), (x_2, y_2), and (x_3, y_3). The area is found using

$$A = \frac{1}{2}\left[\overbrace{(x_1 y_2 - x_2 y_1)}^{A_1} + \overbrace{(x_2 y_3 - x_3 y_2)}^{A_2} + \overbrace{(x_3 y_1 - x_1 y_3)}^{A_3}\right]$$

The centroidal coordinates of the triangle will be at

$$(x_c, y_c) = \left(\frac{x_1 + x_2 + x_3}{3}, \frac{y_1 + y_2 + y_3}{3}\right)$$

The formulas for a triangle can be used to compute properties for the arbitrary n-sided polygon (shown in Figure 4.4) in the xy plane. A polygon can be composed of several triangles with bases defined by the polygon sides. The apex of each polygon lies at the coordinate origin $R_o = (0, 0)$. Some of the triangles will have positive areas while others will have negative areas as illustrated by the six-sided polygon in Figure 4.5.[3] If $R_{i \to j}$ denotes the vector from R_i to R_j, then the triangles with bases defined by $R_{1 \to 2}$, $R_{2 \to 3}$, and $R_{4 \to 5}$ produce positive areas. Similarly, the triangles with bases specified by $R_{3 \to 4}$, $R_{5 \to 6}$, and $R_{6 \to 1}$ produce negative areas. When contributions of all triangles are combined, the original geometry is produced. The polygon is assumed to be traversed in a counterclockwise direction (otherwise the signs will be reversed for positive and negative area contributions).

---

[3] The cross section is defined by the polygon formed by nodes 1,2,3,4,5,6. Triangle O,6,1 provides a negative contribution and triangle O,2,3 provides a postivie contribution to the area summation.

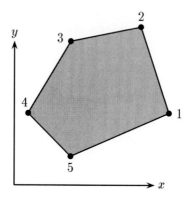

Figure 4.4. General Polygon

Consider a general triangle with corners at R_o, R_i, and R_j for $j = i + 1$ and $1 \leq i \leq n$. When $i = n$, $j = n + 1$ means the corner immediately ahead of corner n (i.e. corner 1). The area and first moment contributions of the triangle are obtained by using results given above. Thus

$$(\text{Area})_i = \frac{1}{2} A_i \qquad A_i = (x_i y_j - x_j y_i)$$

$$(Q_x)_i = \frac{1}{6}(y_i + y_j) A_i$$

$$(Q_y)_i = \frac{1}{6}(x_i + x_j) A_i$$

Computing the inertial moments is slightly more involved but the same ideas apply. The two-dimensional inertia matrix defined by

$$\text{ARR} = \begin{bmatrix} \int_A x^2 \, dA & \int_A xy \, dA \\ \int_A xy \, dA & \int_A y^2 \, dA \end{bmatrix}$$

The inertial property formula for a triangle with corners at R_i, R_j, and $R_o = (0,0)$ simplifies to

$$(\text{ARR})_i = \frac{1}{12} A_i \left[R_i R_i^T + R_j R_j^T + \frac{R_i R_j^T + R_j R_i^T}{2} \right]$$

4.2. Geometrical Properties of Polygons

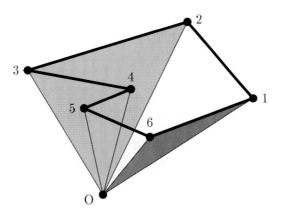

Figure 4.5. General Six-sided Polygon

A typical term $R_i R_j^T$ in this product involves

$$R_i R_j^T = \begin{bmatrix} x_i \\ y_i \end{bmatrix} \begin{bmatrix} x_j & y_j \end{bmatrix} = \begin{bmatrix} x_i x_j & x_i y_j \\ x_j y_i & y_i y_j \end{bmatrix}$$

The equations for geometrical properties of an n-sided polygon, as shown in Figure 4.4, are provided below using a summation of contributions from each side of a polygon created from a set of triangles. An example of a geometry using both positive and negative area contributions is shown in Figure 4.6 and depicts a rectangular cross section with a triangular section removed. One node numbering sequence for this section is $[1, 2, 3, 4, 5, 6, 4, 3, 7]$. The general equations for geometrical properties of a polygon[4] are found to be:

$$A = \frac{1}{2} \sum_{i=1}^{n} A_i, \qquad A_i = (x_i y_j - x_j y_i)$$

$$Q_x = \frac{1}{6} \sum_{i=1}^{n} (y_i + y_j) A_i$$

$$Q_y = \frac{1}{6} \sum_{i=1}^{n} (x_i + x_j) A_i$$

---

[4] A general formulation which treats geometries having curved and straight boundaries can be found in the text by Wilson and Turcotte [54].

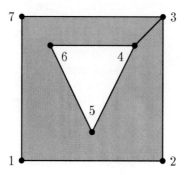

Figure 4.6. Polygon With Positive and Negative Areas

$$I_x = \frac{1}{12} \sum_{i=1}^{n} (y_i^2 + y_i y_j + y_j^2) A_i$$

$$I_y = \frac{1}{12} \sum_{i=1}^{n} (x_i^2 + x_i x_j + x_j^2) A_i$$

$$I_{xy} = \frac{1}{12} \sum_{i=1}^{n} \left[x_i y_i + \frac{x_i y_j + x_j y_i}{2} + x_j y_j \right] A_i$$

$$\bar{x} = \frac{Q_y}{A} \qquad \bar{y} = \frac{Q_x}{A}$$

These concise formulas are used to compute the geometrical properties of a polygon relative to the initial reference axes. The geometrical properties relative to the centroidal axes can then be determined using the parallel-axis theorem.

4.2.1 Program to Determine Geometrical Properties

The equations used to determine the geometrical properties for a polygon are implemented in example program **propex**. Table 4.1 provides a summary of the modules, or functions, included in program **propex** with descriptions of significant

4.2. Geometrical Properties of Polygons

sections of source code within each module. Example program **propex** contains input definitions for several geometries (unit square, channel, partial cross, square with triangular hole, two disconnected triangles, and a 3/4 circle). Only the results from the partial cross geometry are presented here.

Figure 4.7 contains a plot of the geometry for the partial cross. Additionally, the axes for the principal inertial coordinate system are plotted[5]. The axes for the principal inertial system have one line constructed with a dashed line and the remainder are constructed with dotted lines. The dashed line is the reference plane for θ_p (i.e. the angle θ_p is the angle from the positive x-axis to the reference plane for $I_{\max}$, denoted by the dashed line).

Figure 4.8 shows the variation in inertial properties as the centroidal axes are rotated. Several characteristics concerning the relationship between the inertial properties can be observed from this plot. The angle between the planes of $I_{\max}$ and $I_{\min}$ is readily seen to be 90° (i.e., they are perpendicular) and the value for I_{xy} on these planes is zero. The angle between the principal inertial values and the maximum product of inertia is also seen to be 45°. Finally, the plane of maximum I_{xy} occurs when the curves for I_x and I_y intersect (i.e. the average value). The output created by **propex**, which includes the values and angles for principal inertia, is provided in Section 4.2.4.

Table 4.1. Description of Code in Example **propex**

Routine	Line	Operation
propex		script file to execute program.
	22	select one of the example problems.
	23	define the number of values of θ to use for rotation of inertia.
	24-50	define the input parameters.
	54-55	calculate the properties about the original coordinate system.
	57-59	use the parallel-axis theorem to shift the properties to the centroidal axes.
	61-65	calculate the principal inertial values about the centroidal coordinate system.
	68	define a vector of angles at which the rotated inertia values will be computed.
	69-72	calculate the rotated inertia values for the centroidal coordinate systems.
		continued on next page

---

[5] This set of axes is located at the centroid of the geometry.

Routine	Line	Operation
		continued from previous page
	74-121	output the results.
	123-139	define the line segments used for plotting the centroidal and principal inertial axes.
	142	create a closed polygon for plotting.
	143-152	create the plot of the geometry.
	154-164	create the plot of rotated inertia values.
prop		function which calculates the geometrical properties.
	26-36	calculate area, first and second moments of area.
	37-42	if total area is negative user must have numbered nodes backwards, so reverse the signs.
shftprop		shift properties using parallel-axis theorem.
prinert		calculate values and angles for principal inertias.
	42	calculate $I_{\max}$ and $I_{\min}$.
	44-50	define the value which determines zero if user did not pass as argument.
	52-56	handle the special case.
	57-61	handle the general case.
inertang		calculate the inertia values for a specified angle of rotation.

4.2. Geometrical Properties of Polygons

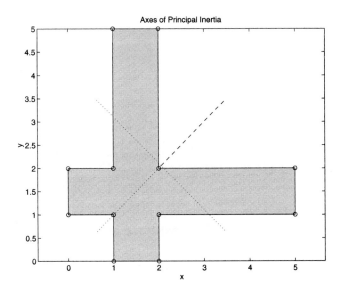

Figure 4.7. Geometry with Principal Axes

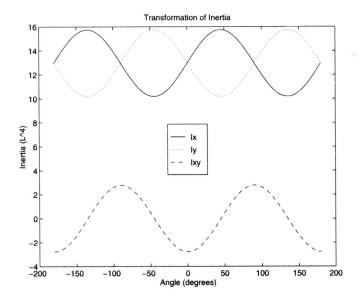

Figure 4.8. Transformation of Inertia

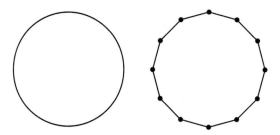

Figure 4.9. Approximation of a Circle Using a Polygon

4.2.2 Using a Polygon to Approximate a Circle

Frequently a cross section will have boundaries defined by arcs, circles, or curves. Polygons can be used quite effectively to approximate these types of boundaries. As an example, program **japprox** demonstrates the use of a polygon as a substitute for a circular cross section. Program **japprox** computes an approximate value for the polar moment of inertia. The polygon is constructed by generating a set of equal chord segments around the circle boundary as shown in Figure 4.9 using function **circle**. These points are connected via line segments, chords, to create the n-sided polygonal representation for a circle. The program compares the approximate value for J to the known value of J ($J = \pi r^4/2$) and increases the number of chord segments used to model the circle until the prescribed allowable error tolerance is achieved. Figure 4.10 shows the relationship between the true value of J and the value found approximately using a polygon. A maximum allowable error of 1% between the two values was used for this analysis. The 1% criterion was achieved by dividing the circle boundary into 37 chords. A 0.1% criteria requires 115 chords and a 0.01% criteria requires 363 chords. This example demonstrates that a polygon can be used effectively to represent complex curved boundaries of a cross section. Table 4.2 provides a summary of the code used to perform this calculation.

4.2.3 Exercises

1. Plot the quantity $I_x + I_y$ as it varies with θ. Can you make any conclusions from this plot?

2. Add the capability to determine the values for maximum and minimum inertia using the fact that these quantities coincide with a reversal of slope in the I versus θ curves.

3. Write a function which will return the value for I_x, I_y, and I_{xy} for a specified value of θ.

4.2. Geometrical Properties of Polygons

4. Write a function to determine the values for inertia about a set of parallel axes using the parallel-axis theorem. How is this routine different from routine **shftprop**?

5. Write a function to determine the radii of gyration for a polygon.

6. Write a function to determine the polar moment of inertia and polar radius of gyration for a polygon.

7. Write a function to determine the polar moment of inertia about a set of parallel axes using the parallel-axis theorem.

8. Write a function to determine the radii of gyration about a set of parallel axes using the parallel-axis theorem.

Table 4.2. Description of Code in Example **japprox**

Routine	Line	Operation
japprox		script file to execute program.
	31	define the input parameters.
	33-34	define some starting values.
	36	calculate true value for J.
	39-52	repeat loop until convergence criteria reached.
	40	output something to let you know program is running.
	48	use function **circle** to generate a set of points which will create the polygon.
	49	calculate the geometrical properties.
	54	calculate the percent error.
	56-66	output the results.
	68-79	plot the results.
prop		function which calculates the geometrical properties.
circle		generate points on a circle. See Appendix A.

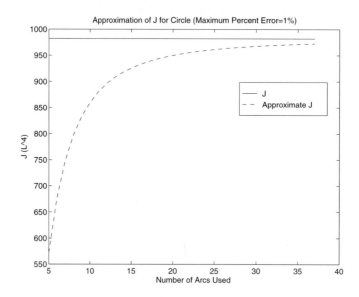

Figure 4.10. Approximation of J for a Circle

4.2.4 Program Output and Code

Output from Example propex

```
    Geometrical Properties
of a Polygonal Cross Section
-----------------------------

Table of Nodes Forming Cross Section

    Node #          x               y
      1             1               0
      2             2               0
      3             2               1
      4             5               1
      5             5               2
      6             2               2
      7             2               5
      8             1               5
      9             1               2
     10             0               2
     11             0               1
```

4.2. Geometrical Properties of Polygons

```
       12              1              1
```

Geometrical Properties
  A: 9
  Qx: 18.5
  Qy: 18.5
  Ix: 51
  Iy: 51
  Ixy: 35.25
  x-bar: 2.05556
  y-bar: 2.05556

Geometrical Properties About Centroidal Axis
  Ix-bar: 12.9722
  Iy-bar: 12.9722
  Ixy-bar: -2.77778

Principal Axes and Inertias for
Centroidal Coordinate System
  I-max: 15.75   at 45 degrees
  I-min: 10.1944   at -45 degrees
  I-avg: 12.9722   at 90 degrees
  Ixy-max: 2.77778   at 90 degrees
  Ixy-min: -2.77778 at 0 degrees
```

**Script File** propex

```
 1: % Example: propex
 2: % ~~~~~~~~~~~~~~~
 3: % This example calculates the geometrical
 4: % properties of a polygonal cross section.
 5: %
 6: % Data is defined in the declaration statements
 7: % below, where:
 8: %
 9: % x,y   - coordinates of polygon corners in
10: %   counterclockwise sequence for
11: %   postive area contributions
12: % No_angs - number of angles to calculate
13: %   inertia values for
14: %
15: % User m functions required:
16: %   prop, shftprop, rotinert, prinert, circle,
```
```

```
17: %     flpang, inertang, genprint
18: %-----------------------------------------
19:
20: clear;
21: %...Input definitions
22: Problem=4;
23: No_angs=90;
24: if Problem == 1
25:    %...Rectangle
26:    x=[-1 1 1 -1]; y=[-5 -5 5 5];
27: elseif Problem == 2
28:    %...Square, all axes are principal
29:    x=[1 1 -1 -1]; y=[-1 1  1  -1];
30: elseif Problem == 3
31:    %...Channel
32:    x=[0  0  1  1  7  7   8 8];
33:    y=[0 -5 -5 -1 -1 -5 -5 0];
34: elseif Problem == 4
35:    %...Partial cross
36:    x=[1 2 2 5 5 2 2 1 0 0 1 ];
37:    y=[0 0 1 1 2 2 5 5 2 2 1 ];
38: elseif Problem == 5
39:    %...Square with triangle cutout
40:    x=[0 10 10 0 0 5 2 8 5];
41:    y=[0 0 10 10 0 2 8 8 2];
42: elseif Problem == 6
43:    %...Two disconnected triangles
44:    x=[5 10 10 5 -10 -5 -10 -10];
45:    y=[0 -6 6 0 -6 0 6 -6];
46: else
47:    %...3/4 circular section
48:    [x,y]=circle(36,0,0,10);
49:    x=x(1:28); x(29)=0; y=y(1:28); y(29)=0;
50: end
51: No_pts=length(x);
52: degrad=pi/180; raddeg=180/pi;
53:
54: %...Find the properties
55: [A,Qx,Qy,Ix,Iy,Ixy,xbar,ybar]=prop(x,y);
56:
57: %...Shift to centroidal axes
58: [Ixbar,Iybar,Ixybar]= ...
59:    shftprop(A,xbar,ybar,Ix,Iy,Ixy);
60:
```

4.2. Geometrical Properties of Polygons

```
61: %...Find principal inertia properties
62: %...for centroidal reference system
63: [Key,Imax,Imin,Iavg,Ixymxmn, ...
64:    Imax_ang,Imin_ang,Ixymax_ang, ...
65:    Ixymin_ang]=prinert(Ixbar,Iybar,Ixybar);
66:
67: %...Find properties for rotated axes
68: theta=-pi:2*pi/No_angs:pi; theta=theta*raddeg;
69: for i=1:No_angs+1
70:    [Ixtheta(i),Iytheta(i),Ixytheta(i)]= ...
71:      inertang(theta(i),Ixbar,Iybar,Ixybar);
72: end
73:
74: %...Output results
75: fprintf('\n\n   Geometrical Properties');
76: fprintf(  '\nof a Polygonal Cross Section');
77: fprintf(  '\n----------------------------');
78: fprintf( ...
79:    '\n\nTable of Nodes Forming Cross Section');
80: fprintf( ...
81:    '\n\n   Node #          x             y');
82: for i=1:No_pts
83:    fprintf('\n  %3.0f  %12.5g  %12.5g', ...
84:          i,x(i),y(i));
85: end
86: fprintf('\n\nGeometrical Properties');
87: fprintf(  '\n   A:    %g',A);
88: fprintf(  '\n   Qx:   %g',Qx);
89: fprintf(  '\n   Qy:   %g',Qy);
90: fprintf(  '\n   Ix:   %g',Ix);
91: fprintf(  '\n   Iy:   %g',Iy);
92: fprintf(  '\n   Ixy:  %g',Ixy);
93: fprintf(  '\n   x-bar: %g',xbar);
94: fprintf(  '\n   y-bar: %g',ybar);
95: fprintf('\n\nGeometrical Properties About ');
96: fprintf(     'Centroidal Axis');
97: fprintf(  '\n   Ix-bar:  %g',Ixbar);
98: fprintf(  '\n   Iy-bar:  %g',Iybar);
99: fprintf(  '\n   Ixy-bar: %g',Ixybar);
100: fprintf('\n\nPrincipal Axes and Inertias for');
101: fprintf(  '\nCentroidal Coordinate System');
102: if Key == 1
103:    fprintf('\n   ------------------------------');
104:    fprintf('\n   ALL axes are principal axes');
```

```
105:    fprintf('\n ------------------------------');
106:    fprintf('\n I-max:         %g',Imax);
107:    fprintf('\n I-min:         %g',Imin);
108:    fprintf('\n Ixy-max/min: %g',Ixymxmn);
109: else
110:    fprintf('\n I-max:         %g',Imax);
111:    fprintf('  at %g degrees',Imax_ang);
112:    fprintf('\n I-min:         %g',Imin);
113:    fprintf('  at %g degrees',Imin_ang);
114:    fprintf('\n I-avg:         %g',Iavg);
115:    fprintf('  at %g degrees',Ixymax_ang);
116:    fprintf('\n Ixy-max:     %g',Ixymxmn);
117:    fprintf('  at %g degrees',Ixymax_ang);
118:    fprintf('\n Ixy-min:     %g',-Ixymxmn);
119:    fprintf('  at %g degrees',Ixymin_ang);
120: end
121: fprintf('\n\n');
122:
123: %...Axes definition for centroid and
124: %...principal centroidal inertia axes plotting
125: xmin=min(x); xmax=max(x); xdiff=xmax-xmin;
126: ymin=min(y); ymax=max(y); ydiff=ymax-ymin;
127: dif=max([xdiff ydiff]); sgn=1;
128: if Imax_ang < 0, sgn=-1; end
129: xl1=xbar-abs(dif/2.5*cos(Imax_ang*degrad));
130: xr1=xbar+abs(dif/2.5*cos(Imax_ang*degrad));
131: yl1=ybar-sgn*abs(dif/2.5*sin(Imax_ang*degrad));
132: yr1=ybar+sgn*abs(dif/2.5*sin(Imax_ang*degrad));
133: xl2=xbar-abs(dif/2.5*sin(Imax_ang*degrad));
134: xr2=xbar+abs(dif/2.5*sin(Imax_ang*degrad));
135: yl2=ybar+sgn*abs(dif/2.5*cos(Imax_ang*degrad));
136: yr2=ybar-sgn*abs(dif/2.5*cos(Imax_ang*degrad));
137: inertx1=[xbar xr1]; inerty1=[ybar yr1];
138: inertx2=[xl1 xbar nan xl2 xr2];
139: inerty2=[yl1 ybar nan yl2 yr2];
140:
141: %...Draw geometry
142: x(No_pts+1)=x(1); y(No_pts+1)=y(1);
143: clf;
144: fill(x,y,'y'); hold on;
145:    plot(x,y,'-',x,y,'o');
146:    plot(inertx1,inerty1,'r--');
147:    plot(inertx2,inerty2,'r:');
148:    xlabel('x'); ylabel('y'); axis('equal');
```

4.2. Geometrical Properties of Polygons

```
149:    title('Axes of Principal Inertia');
150:    hold off; drawnow;
151: %  genprint('geometry');
152:    disp('Press a key to continue'); pause;
153:
154: %...Draw variation of I as rotated
155: %...through 360 degrees
156:    clf;
157:    plot(theta,Ixtheta,'-',theta,Iytheta,':', ...
158:         theta,Ixytheta,'--');
159:    title('Transformation of Inertia');
160:    xlabel('Angle (degrees)');
161:    ylabel('Inertia (L^4)');
162:    tmp=legend('-',' Ix',':',' Iy','--',' Ixy');
163:    axes(tmp); drawnow;
164: %  genprint('transf');
```

Script File japprox

```
 1: % Example: japprox
 2: % ~~~~~~~~~~~~~~~~~
 3: % This program approximates the value for the
 4: % polar moment of inertia, J, for a circular
 5: % cross section.  The circle is approximated
 6: % using an n-sided polygon.
 7: %
 8: % Data is defined in the declaration statements
 9: % below, where:
10: %
11: % radius   - radius of circle with center
12: %            located at (0,0)
13: % epsilon  - allowable error tolerance
14: %            (epsilon*100 gives the maximum
15: %            percent error allowed)
16: % maxarcs - maximum number of arcs to attempt
17: %            to subdivide circle into trying
18: %            to satisfy the error tolerance
19: %
20: % NOTES: Requires-
21: %             37 arcs for max of 1.00% error
22: %            115 arcs for max of 0.10% error
23: %            363 arcs for max of 0.01% error
```

```
24: %
25: % User m functions required:
26: %    circle, genprint
27: %-----------------------------------------
28:
29: clear;
30: %...Input definitions
31: radius=5; epsilon=1e-2; maxarcs=400;
32:
33: Japprox=10e20;   % force thru loop first time
34: No_arcs=4;       % square is first approximation
35:
36: J=pi*radius^4/2; % true value of J
37:
38: i=0;
39: while abs((J-Japprox)/J) > epsilon
40:    fprintf('.');
41:    i=i+1;
42:    if i > maxarcs
43:       fprintf('\n\nMaximum subdivision exceeded');
44:       fprintf('\nArcs attempted: %g\n\n',xp(i-1));
45:       error('Program exit');
46:    end
47:    No_arcs=No_arcs+1;
48:    [x,y]=circle(No_arcs,0,0,radius);
49:    [A,Qx,Qy,Ix,Iy,Ixy,xbar,ybar]=prop(x,y);
50:    Japprox=Ix+Iy; Jcalc(i)=Japprox;
51:    xp(i)=No_arcs;
52: end
53: last=length(xp);
54: PerErr=abs((J-Jcalc(last))/J)*100;
55:
56: fprintf('\n\nApproximation of J');
57: fprintf(   ' for a Circle');
58: fprintf(   '\n------------------');
59: fprintf(   '---------------');
60: fprintf('\n\nEpsilon:        %g',epsilon);
61: fprintf(   ' or %g%%',epsilon*100);
62: fprintf(   '\nArcs required:   %g',xp(last));
63: fprintf(   '\nPercent error:   %g%%',PerErr);
64: fprintf(   '\nTrue J:          %g',J);
65: fprintf(   '\nApproximate J:   %g',Jcalc(last));
66: fprintf('\n');
67:
```

4.2. Geometrical Properties of Polygons

```
68: %...Plot
69: tit=['Approximation of J for Circle',...
70:      ' (Maximum Percent Error=', ...
71:      num2str(epsilon*100),'%)'];
72: clg;
73:   plot([min(xp) max(xp)],[J J],'-');
74:   hold on; plot(xp,Jcalc,'--');
75:   title(tit); xlabel('Number of Arcs Used');
76:   ylabel('J (L^4)');
77:   tmp=legend('-',' J','--',' Approximate J');
78:   axes(tmp); hold off; drawnow;
79: % genprint('geomerr');
```

Function prop

```
 1: function [A,Qx,Qy,Ix,Iy,Ixy,xbar,ybar]= ...
 2:           prop(x,y)
 3: %
 4: % [A,Qx,Qy,Ix,Iy,Ixy,xbar,ybar]=prop(x,y)
 5: % ~~~~~~~~~~~~~~~~~~~~~~~~~~~~~~~~~~~~~~~
 6: % This function computes the geometrical
 7: % properties for a polygon.
 8: %
 9: % x         - vector of x coordinates
10: % y         - vector of x coordinates
11: %
12: % A         - area of polygon
13: % Qx        - first moment of area about
14: %             y-axis
15: % Qy        - first moment of area about
16: %             x-axis
17: % Ix        - inertia about y axis
18: % Iy        - inertia about x axis
19: % Ixy       - product of inertia
20: % xbar      - x distance to centroid
21: % ybar      - y distance to centroid
22: %
23: % User m functions called:  none
24: %------------------------------------------
25:
26: i=1:length(x); j=i+1;
27: A=0; Qx=0; Qy=0; Ix=0; Iy=0; Ixy=0;
```

```
28: x=[x(:);x(1)]; y=[y(:);y(1)];
29: xi=x(i); yi=y(i); xj=x(j); yj=y(j);
30: Ai=xi.*yj-xj.*yi; A=sum(Ai)/2;
31: Qx=sum((yi+yj).*Ai)/6; Qy=sum((xi+xj).*Ai)/6;
32: Ix=sum((yi.^2+yi.*yj+yj.^2).*Ai)/12;
33: Iy=sum((xi.^2+xi.*xj+xj.^2).*Ai)/12;
34: Ixy=sum((xi.*yi+0.5*(xi.*yj+xj.*yi)+xj.*yj) ...
35:        .*Ai)/12;
36: xbar=Qy/A; ybar=Qx/A;
37: %...Reverse signs if necessary (when nodes
38: %   traversed clockwise)
39: if A < 0
40:   A=-A; Qx=-Qx; Qy=-Qy; Ix=-Ix; Iy=-Iy; Ixy=-Ixy;
41: end
```

Function shftprop

```
 1: function [Ixbar,Iybar,Ixybar]=shftprop ...
 2:          (A,xbar,ybar,Ix,Iy,Ixy)
 3: %
 4: % [Ixbar,Iybar,Ixybar]=shftprop ...
 5: %    (A,xbar,ybar,Ix,Iy,Ixy)
 6: %~~~~~~~~~~~~~~~~~~~~~~~~~~~~~~~
 7: % Use parallel-axis theorem to shift properties
 8: % relative to centroidal axis.
 9: %
10: % A       - area of polygon
11: % xbar    - x coordinate of centroid
12: % ybar    - y coordinate of centroid
13: % Ix      - inertia about y axis
14: % Iy      - inertia about x axis
15: % Ixy     - product of inertia
16: %
17: % Ixbar   - inertia about y axis - shifted
18: % Iybar   - inertia about x axis - shifted
19: % Ixybar  - product of inertia - shifted
20: %
21: % User m functions called:   none
22: %----------------------------------------------
23:
24: Ixbar=Ix-A*ybar^2; Iybar=Iy-A*xbar^2;
25: Ixybar=Ixy-A*xbar*ybar;
```

4.2. Geometrical Properties of Polygons

Function prinert

```
 1: function [Key,Imax,Imin,Iavg, ...
 2:    Ixymxmn,Imax_ang,Imin_ang, ...
 3:    Ixymax_ang,Ixymin_ang]= ...
 4:    prinert(Ix,Iy,Ixy,epsilon)
 5: %
 6: % [Key,Imax,Imin,Iavg,Ixymxmn, ...
 7: %    Imax_ang,Imin_ang,Ixymax_ang, ...
 8: %    Ixymin_ang]=prinert(Ix,Iy,Ixy,epsilon)
 9: % ~~~~~~~~~~~~~~~~~~~~~~~~~~~~~~~~~~~~~~~
10: % This function calculates the principal
11: % inertia axes and values.
12: %
13: % Ix          - inertia about y axis
14: % Iy          - inertia about x axis
15: % Ixy         - product of inertia
16: % epsilon     - tolerance for determining if two
17: %               numbers are equal (optional)
18: %
19: % Key         - =0, normal exit
20: %               =1, all axes are principal axes
21: % Imax        - Maximum principal inertia
22: % Imin        - Minimum principal inertia
23: % Iavg        - I for max/min Ixy plane
24: % Ixymxmn     - Max/min for product of inertia
25: % Imax_ang    - Angle for Imax
26: % Imin_ang    - Angle for Imin
27: % Ixymax_ang  - Angle for max Ixymxmn
28: % Ixymin_ang  - Angle for min Ixymxmn
29: %
30: % NOTES:  1) Positive angles measured CCW from
31: %            positive x-axis in degrees.
32: %
33: % User m functions called:  none
34: %-------------------------------------------
35:
36: %...Initialize
37: raddeg=180/pi; Key=0;
38:
39: %...Max/min values
40: Idiff=0.5*(Ix-Iy); Iavg=0.5*(Ix+Iy);
41: Ixymxmn=sqrt(Idiff^2+Ixy^2);
```

```
42:  Imax=Iavg+Ixymxmn; Imin=Iavg-Ixymxmn;
43:
44:  if nargin ~=4
45:    if Imax == Imin
46:      epsilon=1e-3;
47:    else
48:      epsilon=abs(Imax-Imin)*1e-3;
49:    end
50:  end
51:
52:  if abs(Imax-Imin) < epsilon
53:    %...All axes are principal axes
54:    Key=1; Imax_ang=0; Imin_ang=0;
55:    Ixymax_ang=0; Ixymin_ang=0;
56:    Ixymxmn=0; Imax=Ix; Imin=Ix;
57:  else
58:    Imax_ang=raddeg*1/2*atan2(-2*Ixy,Ix-Iy);
59:    Imin_ang=Imax_ang+90; Ixymax_ang=Imax_ang+45;
60:    Ixymin_ang=Imax_ang-45;
61:  end
```

Function inertang

```
 1:  function [Ix_angle,Iy_angle,Ixy_angle]= ...
 2:           inertang(Angle,Ix,Iy,Ixy)
 3:  %
 4:  % [Ix_angle,Iy_angle,Ixy_angle]= ...
 5:  %    inertang(Angle,Ix,Iy,Ixy)
 6:  % ~~~~~~~~~~~~~~~~~~~~~~~~~~~~~~~~
 7:  % This function calculates the inertia
 8:  % values for a geometry rotated by
 9:  % angle Angle.
10:  %
11:  % Angle     - angle to determine the inertia
12:  %             values for in degrees.
13:  %             (positive measured CCW from
14:  %             positive x-axis)
15:  % Ix        - inertia about y-axis
16:  % Iy        - inertia about x-axis
17:  % Ixy       - product of inertia about x-y
18:  %
19:  % Ix_angle  - Ix at Angle
```

4.2. Geometrical Properties of Polygons

```
20: % Iy_angle  - Iy at Angle
21: % Ixy_angle - Ixy at Angle
22: %
23: % User m functions called:  none
24: %-------------------------------------------
25:
26: degrad=pi/180;
27: angle_rad=degrad*Angle; t2=2*angle_rad;
28: Iavg=0.5*(Ix+Iy); Idiff=0.5*(Ix-Iy);
29: cost2=cos(t2); sint2=sin(t2);
30: Ix_angle=Iavg+Idiff*cost2-Ixy*sint2;
31: Iy_angle=Iavg-Idiff*cost2+Ixy*sint2;
32: Ixy_angle=Idiff*sint2+Ixy*cost2;
```

Chapter 5

Transformation of Stress and Strain

The transformation of stress and strain is one of the most fundamental topics in mechanics of materials. In this chapter the stresses and strains which act on differential elements are studied. The relationships for stresses and strains acting on rotated planes are presented. A program is included for the analysis of plane stress[1]. The development of the analogous functions for plane strain are left as an exercise for the reader. Another program is presented for analyzing the experimental results produced by using strain rosettes to measure strains in a member subjected to loading. Strain rosettes are commonly employed for laboratory measurements and the calculation of the corresponding stresses can be efficiently implemented using a simple computer program.

5.1 Transformation of Plane Stress

Figure 5.1 depicts a plane stress element with its corresponding stresses: the normal stresses σ_x and σ_y and the shear stresses τ_{xy}. The element shown in Figure 5.1 depicts the sign convention which will be used for plane stress. Normal stresses are defined as positive when they induce tension. Shear stresses are defined as positive when the shear stress on the right vertical face (σ_x is also on this face) is directed downward. If this face is rotated the sign convention is simply rotated with the element. The reader should be reminded that the shear stresses always occur as two sets of opposing couples. This is the result of the moment equilibrium requirement for the element. Therefore, once the direction of one shear stress is known the other three are known.

The stresses on the element shown in Figure 5.1 do not necessarily represent the extreme magnitudes of stress for that particular point in a member. Stresses on planes at an angle of inclination with this element, as shown in Figure 5.2, can have

---

[1] The relationships developed for the transformation of plane stress and plane strain are mathematically analogous to those developed previously in Chapter 4 for transforming geometrical properties.

5.1. Transformation of Plane Stress

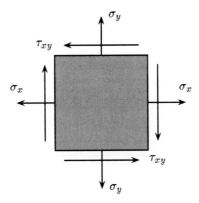

Figure 5.1. Sign Convention for Plane Stress

values of stress exceeding those on the original element. The analysis and design of members requires that the maximum and minimum stresses, or *principal stresses*, be determined. The remainder of this section presents the relationships necessary to determine the principal stresses and the angle of inclination for these stresses relative to the original element.

5.1.1 Rotation of Plane Stress

The stress values are dependent on the set of axes chosen as shown in Figure 5.2. Therefore, the stress values will change as the perpendicular reference axes are rotated. The relationships for the stress values when rotated through an angle θ (θ is measured from the positive x-axis and is positive in the counterclockwise direction) are

$$\sigma_\theta = \frac{\sigma_x + \sigma_y}{2} + \frac{\sigma_x - \sigma_y}{2}\cos(2\theta) - \tau_{xy}\sin(2\theta)$$

$$\sigma_{\theta+90°} = \frac{\sigma_x + \sigma_y}{2} - \frac{\sigma_x - \sigma_y}{2}\cos(2\theta) + \tau_{xy}\sin(2\theta)$$

$$\tau_\theta = \frac{\sigma_x - \sigma_y}{2}\sin(2\theta) + \tau_{xy}\cos(2\theta)$$

The values for the maximum and minimum stress may be found from these relationships by differentiating σ with respect to θ and setting the resulting equation

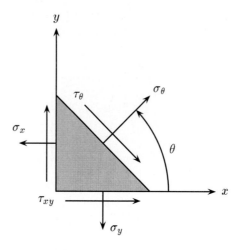

Figure 5.2. Stress on Rotated Plane

equal to zero. The two planes obtained from this equation are called the *principal axes of stress* and are located by

$$\tan(2\theta_p) = \frac{-2\tau_{xy}}{\sigma_x - \sigma_y}$$

where the angle θ_p is measured from the positive x-axis to the principal planes. The above equation has two solutions for θ_p. One solution produces the angle to $\sigma_{\max}$ and the other solution produces the angle to $\sigma_{\min}$. Note that if $\tau_{xy} \neq 0$ and $\sigma_x = \sigma_y$ implies $\theta_p = \pm\pi/4$. The reader should note that the sign for the numerator has been maintained as part of the numerator rather than negating the entire equation (as many texts present this equation). Writing the equation in this fashion allows the value of θ_p to be associated with the proper principal inertial value[2]. The values for stress on the principal planes can be summarized as

$$\sigma_{\max} = \frac{\sigma_x + \sigma_y}{2} + \sqrt{\left(\frac{\sigma_x - \sigma_y}{2}\right)^2 + (\tau_{xy})^2}$$

$$\sigma_{\min} = \frac{\sigma_x + \sigma_y}{2} - \sqrt{\left(\frac{\sigma_x - \sigma_y}{2}\right)^2 + (\tau_{xy})^2}$$

---

[2] The quadrant in which this angle lies can be determined by the signs of the numerator and denominator.

5.1. Transformation of Plane Stress

where $\sigma_{\max}$ and $\sigma_{\min}$ are perpendicular and 90° apart and τ_{xy} on the principal planes is zero. The maximum and minimum value for τ_{xy} is produced on a plane 45° from the planes for $\sigma_{\max}$ and $\sigma_{\min}$, or

$$(\tau_{xy})_{\max/\min} = \pm\sqrt{\left(\frac{\sigma_x - \sigma_y}{2}\right)^2 + (\tau_{xy})^2}$$

and σ_x and σ_y on these planes is non-zero. The equation for the angle to the plane of maximum and minimum τ_{xy} is

$$\tan(2\alpha) = \frac{\sigma_x - \sigma_y}{2\tau_{xy}}$$

However, in practice this equation is rarely used since the $(\tau_{xy})_{\max}$ plane is +45° from the plane of $\sigma_{\max}$ and the plane to $(\tau_{xy})_{\min}$ plane is −45° from the plane of $\sigma_{\max}$. Finally, if the denominator of the above equation is zero (i.e. $\tau_{xy} = 0$) then the present axes are principal stress axes.

5.1.2 Program for Plane Stress

The equations used to determine the stresses occurring on an specified angle of inclination are implemented in example program **strglex**. The equations for determining the principal stresses and their associated angles of inclination are implemented in example program **prstrex**. The output produced from both of this programs is trivial and is not included. However, the plots created by these programs contain all the results in a more comprehensive graphical form. A summary of the computer code for both of these programs is provided in Table 5.1.

Figure 5.3 provides the results from executing program **strglex** to determine the stresses on a prescribed angle of inclination. The plot shows both the original element and the element rotated through the angle. A dotted line is used to indicate the rotation of the axis of reference between the original element and the rotated element. Figure 5.4 depicts the results produced by executing program **prstrex**. Once again a dotted line is used to provide a reference between the original element and the rotated elements.

The distribution of stresses as the element is rotated through 360° can be studied quite effectively by simply plotting the corresponding values for normal and shear stresses. Figure 5.5 shows the normal and shear stress values for each possible plane of rotation. Inspection of this graph provides a wealth of information: a) the angle between a maximum and minimum value of stress is always 90°; b) the angle between the maximum normal stress and the maximum shear stress is always 45°; c) the value for the shear stress on the planes of principal normal stress is zero; d) the average value for the normal stress occurs where the curves for the two normal stresses intersect and the shear stress on this plane is a maximum; e) the average value for the shear stress is zero; f) the plane producing a maximum normal stress also produces a minimum normal stress and vice versa; and g) the curves have a period of π. An alternative method of graphing the distribution of stresses is shown in Figure 5.6. This figure plots the relationship between angle of inclination and stress using a polar diagram. The reader should study this diagram to understand the effectiveness of using a polar representation to present this information.

5.1.3 Exercises

1. Plot the quantity $\sigma_x + \sigma_y$ as it varies with θ. Can you make any conclusions from this plot?

Table 5.1. Description of Code in Examples **strglex** & **prstrex**

Routine	Line	Operation
strglex		script file to execute program for determining the stresses on a prescribed angle of inclination.
	28	select one of the example problems.
	29-32	define the input parameters.
	34-36	use function **strangle** to determine the stresses on the plane.
	38-50	output the results.
	53	make the angle be in first or fourth quadrants.
	54-55	create a plot of results using function **strangpl**.
prstrex		script file to execute program for determining the principal stresses for an element.
	30	select one of the example problems.
	32-38	define the input parameters.
	41-44	call function **prstress** to determine the principal stresses.
	47	define a vector of angles from $-\pi$ to $+\pi$.
	48-51	determine the stresses for each angle using function **strangle**.
	53-80	output the results.
		continued on next page

5.1. Transformation of Plane Stress

Routine	Line	Operation
		continued from previous page
	82-86	create a plot of results using function **prplot**.
	88-97	create the plot of stress versus angle.
	99-109	create the polar plot of stress versus angle using function **polarstr**.
strangle		function used to determine the stresses corresponding to a specified angle of inclination.
prstress		function used to determine the principal stresses and their associated angles.
	46	calculate the average normal stress.
	48	calculate the maximum/minimum shear stresses.
	49	calculate the principal normal stresses.
	51-57	define the numerical measure of zero if not specified in call to function **prstress**.
	59-64	hydrostatic case and all axes are principal.
	65-73	already in state of maximum or minimum shear stress.
	74-84	already in state of principal normal stress.
	85-97	calculate the angles to the principal stresses.
	87	use intrinsic function **atan2** to find the angle to σ_{max}.
	89-91	calculate the other principal angles.
	93-97	make all angle measures be in the first and fourth quadrant using function **flpang**.
strangpl		function used to plot the stresses on a specified plane of inclination.
	33	define the unit element for plotting.
	34	define the coordinates for the dotted line to be used as reference line.
		continued on next page

Routine	Line	Operation
		continued from previous page
	43-64	create the plot of the original element.
	40-41	define the directions for the arrowheads on the stress vectors.
	43	save the graphics identifier, **ax**, for this figure.
	47-50	plot the vectors for σ_x using function **arrows**.
	51-54	plot the vectors for σ_y using function **arrows**.
	55-60	plot the vectors for τ_{xy} using function **arrows**.
	64	use undocumented MATLAB feature to save the range of the axes.
	69-107	create the plot of the inclined element.
	80-81	define the standard rotation matrix.
	82-83	perform the rotation transformation for the element and reference axis.
	105-106	force the axes to have the same range as the plot of the original element.
prplot		function used to plot the stresses on the principal planes.
arrows		function used to plot the stress vectors with their appropriate arrowheads.
polarstr		function used to create the polar plot of stress versus angle.
flpang		make the angles always be in first and fourth quadrants. (See Appendix A.)

5.1. Transformation of Plane Stress

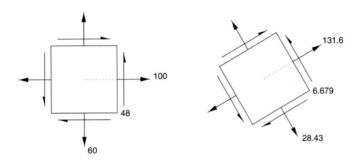

Figure 5.3. Stresses on Inclined Plane

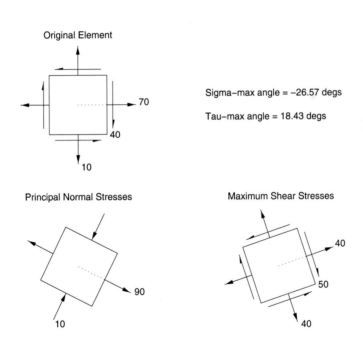

Figure 5.4. Principal Stresses

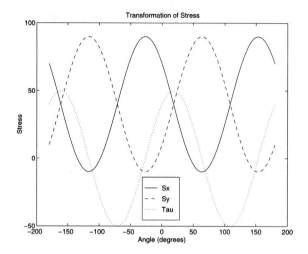

Figure 5.5. Transformation of Stress

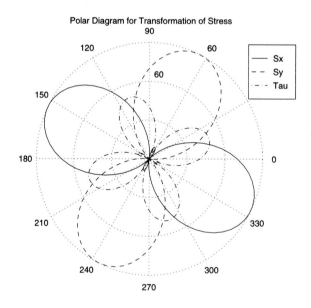

Figure 5.6. Polar Diagram for Transformation of Stress

5.1.4 Program Output and Code

Script File strglex

```
 1: % Example: strglex
 2: % ~~~~~~~~~~~~~~~~
 3: % This example determines the stresses on
 4: % a specified plane.
 5: %
 6: % Data is defined in the declaration statements
 7: % below, where:
 8: %
 9: % Sigma_x  - normal stress on x-axis
10: % Sigma_y  - normal stress on y-axis
11: % Tau_xy   - shear stress
12: % Angle    - angle (in degrees) to determine
13: %            stresses for
14: %
15: % NOTES: 1) normal stresses are + for tension,
16: %           - for compression
17: %        2) shear stresses are + when they
18: %           are directed down on the positive
19: %           x-axis face
20: %
21: % User m functions required:
22: %    strangle, strangpl, arrows, flpang,
23: %    genprint
24: %-----------------------------------------------
25:
26: clear;
27: %...Problem definition
28: Problem=1;
29: if Problem == 1
30:    Sigma_x=100; Sigma_y=60; Tau_xy=-48;
31:    Angle=30;
32: end
33:
34: %...Find stresses on the plane
35: [Sx_ang,Sy_ang,Tau_ang]=strangle( ...
36:    Angle,Sigma_x,Sigma_y,Tau_xy);
37:
38: fprintf ...
39:    ('\n\nDetermination of Stresses on a Plane');
```

```
40: fprintf ...
41:    ('\n----------------------------------\n');
42: fprintf('\n  Sigma_x: %g',Sigma_x);
43: fprintf('\n  Sigma_y: %g',Sigma_y);
44: fprintf('\n  Tau_xy:  %g\n',Tau_xy);
45: fprintf('\n  Sigma: %g at %g', ...
46:    Sx_ang,Angle);
47: fprintf('\n  Sigma: %g at %g', ...
48:    Sy_ang,Angle+90);
49: fprintf('\n  Tau:   %g at %g\n', ...
50:     Tau_ang,Angle);
51:
52: %...Plot results
53: [NewAngle]=flpang(Angle,0);
54: strangpl(Sigma_x,Sigma_y,Tau_xy, ...
55:    NewAngle,Sx_ang,Sy_ang,Tau_ang);
56: %genprint('strangle')
```

Script File prstrex

```
 1: % Example: prstrex
 2: % ~~~~~~~~~~~~~~~~
 3: % This example determines the principal
 4: % stresses for an element.
 5: %
 6: % Data is defined in the declaration statements
 7: % below, where:
 8: %
 9: % No_angs - number of angles to calculate
10: %           stress values for
11: % Sigma_x - normal stress on x-axis
12: % Sigma_y - normal stress on y-axis
13: % Tau_xy  - shear stress
14: % Angle   - angle (in degrees) to determine
15: %           stresses for
16: %
17: % NOTES: 1) normal stresses are + for tension,
18: %           - for compression
19: %        2) shear stresses are + when they
20: %           are directed down on the positive
21: %           x-axis face
22: %
```

5.1. Transformation of Plane Stress

```
23: % User m functions required:
24: %    prstress, strangle, prplot, arrows,
25: %    polarstr, genprint
26: %-------------------------------------------
27:
28: clear;
29: %...Input definitions
30: Problem=2;
31: No_angs=90;
32: if Problem == 1
33:   %...Hydrostatic case
34:   Sigma_x=10; Sigma_y=10; Tau_xy=0;
35: elseif Problem == 2
36:   %...Other
37:   Sigma_x=70; Sigma_y=10; Tau_xy=40;
38: end
39: degrad=pi/180; raddeg=180/pi;
40:
41: %...Find principal stresses
42: [Key,Smax,Smin,Savg,Tmxmn,Smax_ang, ...
43:    Smin_ang,Tmax_ang,Tmin_ang]= ...
44:    prstress(Sigma_x,Sigma_y,Tau_xy);
45:
46: %...Find properties for rotated axes
47: theta=linspace(-180,180,No_angs+1);
48: for i=1:No_angs+1
49:   [Sxtheta(i),Sytheta(i),Sxytheta(i)]= ...
50:      strangle(theta(i),Sigma_x,Sigma_y,Tau_xy);
51: end
52:
53: %...Output results
54: fprintf ...
55:   ('\n\nDetermination of Principal Stresses');
56: fprintf ...
57:   ('\n---------------------------------\n');
58: fprintf('\n   Sigma-x: %g',Sigma_x);
59: fprintf('\n   Sigma-y: %g',Sigma_y);
60: fprintf('\n   Tau-xy:  %g\n',Tau_xy);
61: if Key == 1
62:   fprintf('\n   -----------------------------');
63:   fprintf('\n   ALL axes are principal axes');
64:   fprintf('\n   -----------------------------');
65:   fprintf('\n   Sigma-max: %g',Smax);
66:   fprintf('\n   Sigma-min: %g',Smin);
```

```
67:      fprintf('\n  Tau-max:    %g',Tmxmn);
68: else
69:      fprintf('\n  Sigma-max: %g at %g', ...
70:        Smax,Smax_ang);
71:      fprintf('\n  Sigma-min: %g at %g', ...
72:        Smin,Smin_ang);
73:      fprintf('\n  Sigma-avg: %g at %g', ...
74:        Savg,Tmax_ang);
75:      fprintf('\n  Tau-max:   %g at %g', ...
76:        Tmxmn,Tmax_ang);
77:      fprintf('\n  Tau-min:   %g at %g', ...
78:        -Tmxmn,Tmin_ang);
79: end
80: fprintf('\n\n');
81:
82: %...Plot the results
83: prplot(Smax,Smin,Savg,Tmxmn,Smax_ang, ...
84:     Tmax_ang,Sigma_x,Sigma_y,Tau_xy)
85: % genprint('prstress');
86:    disp('Press a key to continue'); pause;
87:
88: clf;
89: plot(theta,Sxtheta,'-',theta,Sytheta,'--', ...
90:       theta,Sxytheta,':');
91:    title('Transformation of Stress');
92:    xlabel('Angle (degrees)');
93:    ylabel('Stress');
94:    h=legend('-',' Sx','--',' Sy',':',' Tau');
95:    axes(h); drawnow;
96: % genprint('rotstr')
97:    disp('Press a key to continue'); pause;
98:
99: tit=['Polar Diagram for Transformation ', ...
100:      'of Stress'];
101: theta=theta*degrad;
102: clf;
103:    polarstr(theta,Sxtheta,'w-'); hold on;
104:    polarstr(theta,Sytheta,'c--');
105:    polarstr(theta,Sxytheta,'g-.');
106:    title(tit); axis('equal'); axis('off');
107:    h=legend('w-',' Sx','c--',' Sy','g-.',' Tau');
108:    hold off; axes(h); drawnow;
```

5.1. Transformation of Plane Stress

109: `% genprint('polar')`

Function strangle

```
 1: function [Sx_angle,Sy_angle,Tau_angle]= ...
 2:          strangle(Angle,Sigma_x,Sigma_y,Tau_xy)
 3: %
 4: % [Sx_angle,Sy_angle,Tau_angle]= ...
 5: %          strangle(Angle,Sigma_x,Sigma_y,Tau_xy)
 6: %~~~~~~~~~~~~~~~~~~~~~~~~~~~~~~~~~~~~~~~~~~~
 7: % This function calculates the normal and
 8: % shear stresses for an element rotated by
 9: % angle Angle.
10: %
11: % Angle      - angle to determine the stress
12: %              values for in degrees.
13: %              (positive measured CCW from
14: %              positive x-axis)
15: % Sigma_x    - sigma in the x direction
16: %              (positive in tension)
17: % Sigma_y    - sigma in the y direction
18: %              (positive in tension)
19: % Tau_xy     - tau in the xy plane
20: %              (positive is up on right face)
21: %
22: % Sx_angle   - sigma_x at Angle
23: % Sy_angle   - sigma_y at Angle
24: % Tau_angle  - tau at Angle
25: %
26: % User m functions called:  none
27: %-------------------------------------------
28:
29: degrad=pi/180;
30: angle_rad=Angle*degrad; t2=2*angle_rad;
31: Savg=(Sigma_x+Sigma_y)/2;
32: Sdiff=(Sigma_x-Sigma_y)/2;
33: cost2=cos(t2); sint2=sin(t2);
34: Sx_angle=Savg+Sdiff*cost2-Tau_xy*sint2;
35: Sy_angle=Savg-Sdiff*cost2+Tau_xy*sint2;
36: Tau_angle=Sdiff*sint2+Tau_xy*cost2;
```

Function prstress

```
 1: function [Key,Smax,Smin,Savg,Tmxmn, ...
 2:    Smax_ang,Smin_ang,Tmax_ang,Tmin_ang]= ...
 3:    prstress(Sigma_x,Sigma_y,Tau_xy,epsilon)
 4: %
 5: % [Key,Smax,Smin,Savg,Tmxmn,Smax_ang, ...
 6: %   Smin_ang,Tmax_ang,Tmin_ang]= ...
 7: %   prstress(Sigma_x,Sigma_y,Tau_xy,epsilon)
 8: %~~~~~~~~~~~~~~~~~~~~~~~~~~~~~~~~~~~~~~~~~
 9: % This function calculates the principal normal
10: % and shear stresses for an element.
11: %
12: % Sigma_x   - sigma in the x direction
13: % Sigma_y   - sigma in the y direction
14: % Tau_xy    - tau in the xy plane
15: % epsilon   - tolerance for determining
16: %             if two numbers are equal
17: %             (optional)
18: %
19: % Key       - =0, normal exit
20: %             =1, all axes are principal
21: % Smax      - Sigma maximum
22: % Smin      - Sigma minimum
23: % Savg      - Sigma on planes of max/min
24: %             shear stress
25: % Tmxmn     - Tau maximum. Tau minimum is
26: %             negative of Tau maximum
27: % Smax_ang - angle to Sigma_max
28: % Smin_ang - angle to Sigma_min
29: % Tmax_ang - angle to Tau_max
30: % Tmin_ang - angle to Tau_min
31: %
32: % NOTES:  1)  All angles measured CCW from
33: %             positive x-axis in degrees.
34: %         2)  Normal stresses are + for
35: %             tension, - for compression.
36: %         3)  Shear stresses are + for
37: %             up on right face.
38: %
39: % User m functions called:  strangle
40: %-----------------------------------------
41:
```

5.1. Transformation of Plane Stress

```
42:  %...Initialize
43:  raddeg=180/pi; Key=0;
44:
45:  %...Max/min values
46:  Savg=(Sigma_x+Sigma_y)/2;
47:  Sdiff=(Sigma_x-Sigma_y)/2;
48:  Tmxmn=sqrt(Sdiff^2+Tau_xy^2);
49:  Smax=Savg+Tmxmn; Smin=Savg-Tmxmn;
50:
51:  if nargin ~= 4
52:    if Smax == Smin
53:      epsilon=1e-3;
54:    else
55:      epsilon=abs(Smax-Smin)*1e-3;
56:    end
57:  end
58:
59:  if abs(Smax-Smin) < epsilon
60:    %...Hydrostate case, all axes are principal
61:    Key=1;
62:    Smax=Sigma_x; Smin=Sigma_x; Tmxmn=0;
63:    Smax_ang=0; Smin_ang=0;
64:    Tmax_ang=0; Tmin_ang=0;
65:  elseif abs(Sigma_x-Sigma_y) < epsilon
66:    %...In state of max/min shear
67:    if Tau_xy < 0
68:      Smax_ang=45; Smin_ang=-45;
69:      Tmax_ang=90; Tmin_ang=0;
70:    else
71:      Smax_ang=-45; Smin_ang=45;
72:      Tmax_ang=0; Tmin_ang=90;
73:    end
74:  elseif abs(Tau_xy) < epsilon
75:    %...In state of max/min normal stress
76:    if Sigma_x > Sigma_y
77:      %...sigma-x is max
78:      Smax_ang=0; Smin_ang=90;
79:      Tmax_ang=45; Tmin_ang=-45;
80:    else
81:      %...sigma-y is max
82:      Smax_ang=90; Smin_ang=0;
83:      Tmax_ang=-45; Tmin_ang=45;
84:    end
85:  else
```

```
86:     %...Get angle to Smax plane
87:     angle=0.5*atan2(-2*Tau_xy,Sigma_x-Sigma_y);
88:     Smax_ang=angle*raddeg;
89:     Smin_ang=(angle+pi/2)*raddeg;
90:     Tmax_ang=Smax_ang+45;
91:     Tmin_ang=Tmax_ang+90;
92:
93:     %...Make angles between -90 and +90
94:     Smax_ang=flpang(Smax_ang,0);
95:     Smin_ang=flpang(Smin_ang,0);
96:     Tmax_ang=flpang(Tmax_ang,0);
97:     Tmin_ang=flpang(Tmin_ang,0);
98: end
```

Function strangpl

```
 1: function strangpl(Sigma_x,Sigma_y,Tau_xy, ...
 2:     Angle,Sigma_ang0,Sigma_ang90,Tau_ang)
 3: %----------------------------------------------
 4: %
 5: % strangpl(Sigma_x,Sigma_y,Tau_xy, ...
 6: %     Angle,Sigma_ang0,Sigma_ang90,Tau_ang)
 7: % ~~~~~~~~~~~~~~~~~~~~~~~~~~~~~~~~~~~~~~~
 8: % This function plots the element and the
 9: % stresses on a rotated element.
10: %
11: % Sigma_x      - sigma in the x direction
12: % Sigma_y      - sigma in the y direction
13: % Tau_xy       - tau in the xy plane
14: % Angle        - angle to stress plane (degs)
15: % Sigma_ang0   - sigma at Angle
16: % Sigma_ang90  - sigma at Angle+90degs
17: % Tau_ang      - tau at Angle
18: %
19: % NOTE: To make the plotting perform
20: %       as desired an undocumented MATLAB
21: %       capability has been used (Renderlimits).
22: %       MathWorks does not promise that it
23: %       will be supported in the future.
24: %       Renderlimits allows the two side-by-
25: %       side plots to maintain the same
26: %       y-axes.
```

```
27: %
28: % User m functions called:  arrows
29: %--------------------------------------------
30:
31: degrad=pi/180; pi2=pi/2;
32: %...Define a unit square for element
33: Element=[-1 -1; 1 -1; 1 1; -1 1; -1 -1];
34: Raxis=[0 0;1 0];
35:
36: %...................
37: %...Original element
38: %...................
39: %...Arrowhead direction
40: dirSx=sign(Sigma_x); dirSy=sign(Sigma_y);
41: dirTxy=sign(Tau_xy);
42:
43: clf; ax=subplot(1,2,1);
44:    plot(Element(:,1),Element(:,2),'-');
45:      hold on;
46:      plot(Raxis(:,1),Raxis(:,2),':');
47:      if Sigma_x ~= 0
48:         arrows(dirSx,0,0,Sigma_x,0,0,1,0);
49:         arrows(dirSx,0,pi,0,0,0,0,0);
50:      end
51:      if Sigma_y ~= 0
52:         arrows(dirSy,0,pi2,0,0,0,0,0);
53:         arrows(dirSy,0,-pi2,0,Sigma_y,0,2,0);
54:      end
55:      if Tau_xy ~= 0
56:         arrows(-dirTxy,-1,0,0,0,Tau_xy,3,Element);
57:         arrows(-dirTxy,-1,pi,0,0,0,0,0);
58:         arrows(dirTxy,-1,pi2,0,0,0,0,0);
59:         arrows(dirTxy,-1,-pi2,0,0,0,0,0);
60:      end
61:      title('Original Element');
62:      axis('equal'); axis('off');
63:      hold off; drawnow;
64:      Rlimits=get(ax,'RenderLimits');
65:
66: %................
67: %...Stress plane
68: %................
69: angle=Angle*degrad;
70: S1=Sigma_ang0; S2=Sigma_ang90;
```

```
71:  T=Tau_ang;
72:  str1=['Stresses at angle = ', ...
73:        num2str(Angle),' degs'];
74:
75:  %...Arrowhead direction
76:  dirSx=sign(Sigma_ang0); dirSy=sign(Sigma_ang90);
77:  dirTxy=sign(Tau_xy);
78:
79:  %...Rotate by angle
80:  trans=[cos(angle) -sin(angle);
81:         sin(angle)  cos(angle)];
82:  temp=trans*Element'; Elementt=temp';
83:  temp=trans*Raxis'; Raxist=temp';
84:
85:  subplot(1,2,2);
86:    plot(Elementt(:,1),Elementt(:,2),'-');
87:    hold on;
88:    plot(Raxist(:,1),Raxist(:,2),':');
89:    if S1 ~= 0
90:       arrows(dirSx,0,angle,S1,0,0,1,0);
91:       arrows(dirSx,0,pi+angle,0,0,0,0,0);
92:    end
93:    if S2 ~= 0
94:       arrows(dirSy,0,pi2+angle,0,0,0,0,0);
95:       arrows(dirSy,0,-pi2+angle,0,S2,0,2,0);
96:    end
97:    if T ~= 0
98:       arrows(-dirTxy,-1,angle,0,0,0,0,0);
99:       arrows(-dirTxy,-1,pi+angle,0,0,T,3, ...
100:              Elementt);
101:       arrows(dirTxy,-1,pi2+angle,0,0,0,0,0);
102:       arrows(dirTxy,-1,-pi2+angle,0,0,0,0,0);
103:    end
104:    title(str1); axis('off');
105:    set(gca,'Xlim',[Rlimits(1) Rlimits(2)]);
106:    set(gca,'Ylim',[Rlimits(3) Rlimits(4)]);
107:    hold off; drawnow;
```

Function prplot

```
1:  function prplot(Sigma_max,Sigma_min, ...
2:    Sigma_avg,Tau_mxmn,Sigma_max_ang, ...
```

5.1. Transformation of Plane Stress

```
 3:      Tau_max_ang,Sigma_x,Sigma_y,Tau_xy)
 4: %
 5: % prplot(Sigma_max,Sigma_min, ...
 6: %    Sigma_avg,Tau_mxmn,Sigma_max_ang, ...
 7: %    Tau_max_ang,Sigma_x,Sigma_y,Tau_xy)
 8: %~~~~~~~~~~~~~~~~~~~~~~~~~~~~~~~~~~~~~~
 9: % This function plots the element, the
10: % principal normal stresses and the shear
11: % stresses for an element.
12: %
13: % Sigma_max      - Sigma maximum
14: % Sigma_min      - Sigma minimum
15: % Sigma_avg      - Sigma on planes of max/min
16: %                  shear stress
17: % Tau_mxmn       - Tau maximum, Tau minimum is
18: %                  negative of Tau maximum
19: % Sigma_max_ang - angle to Sigma_max (degs)
20: % Tau_max_ang   - angle to Tau_max (degs)
21: % Sigma_x        - sigma in the x direction
22: % Sigma_y        - sigma in the y direction
23: % Tau_xy         - tau in the xy plane
24: %
25: % NOTE: To make the plotting perform
26: %       as desired an undocumented MATLAB
27: %       capability has been used (Renderlimits).
28: %       MathWorks does not promise that it
29: %       will be supported in the future.
30: %       Renderlimits allows the two side-by-
31: %       side plots to maintain the same
32: %       y-axes.
33: %
34: % User m functions called:   arrows
35: %-----------------------------------------
36:
37: degrad=pi/180; pi2=pi/2;
38: %...Define a unit square for element
39: Element=[-1 -1; 1 -1; 1 1; -1 1; -1 -1];
40: Raxis=[0 0;1 0];
41:
42: %....................
43: %...Original element
44: %....................
45: %...Arrowhead direction
46: dirSx=sign(Sigma_x); dirSy=sign(Sigma_y);
```

```
47: dirTxy=sign(Tau_xy);
48:
49: clf; ax=subplot(2,2,1);
50:    plot(Element(:,1),Element(:,2),'-');
51:      hold on;
52:      plot(Raxis(:,1),Raxis(:,2),':');
53:      if Sigma_x ~= 0
54:         arrows(dirSx,0,0,Sigma_x,0,0,1,0);
55:         arrows(dirSx,0,pi,0,0,0,0,0);
56:      end
57:      if Sigma_y ~= 0
58:         arrows(dirSy,0,pi2,0,0,0,0,0);
59:         arrows(dirSy,0,-pi2,0,Sigma_y,0,2,0);
60:      end
61:      if Tau_xy ~= 0
62:         arrows(-dirTxy,-1,0,0,0,Tau_xy,3,Element);
63:         arrows(-dirTxy,-1,pi,0,0,0,0,0);
64:         arrows(dirTxy,-1,pi2,0,0,0,0,0);
65:         arrows(dirTxy,-1,-pi2,0,0,0,0,0);
66:      end
67:      title('Original Element');
68:      axis('equal'); axis('off');
69:      hold off; drawnow;
70:      Rlimits=get(ax,'Renderlimits');
71:
72: %..............
73: %...Text angles
74: %..............
75: str1=['Sigma-max angle = ', ...
76:       num2str(Sigma_max_ang),' degs'];
77: str2=['Tau-max angle = ', ...
78:       num2str(Tau_max_ang),' degs'];
79: subplot(2,2,2);
80:    axis([0 1 0 1]); axis('off');
81:    text(0.0,0.6,str1);
82:    text(0.0,0.4,str2);
83:    drawnow;
84:
85: %.....................
86: %...Principal stresses
87: %.....................
88: angle=Sigma_max_ang*degrad;
89: S1=Sigma_max; S2=Sigma_min;
90:
```

5.1. Transformation of Plane Stress

```
91:  %...Arrowhead direction
92:  dirSx=sign(Sigma_max); dirSy=sign(Sigma_min);
93:
94:  %...Rotate by angle
95:  trans=[cos(angle) -sin(angle);
96:         sin(angle)  cos(angle)];
97:  temp=trans*Element'; Elementt=temp';
98:  temp=trans*Raxis';   Raxist=temp';
99:
100: subplot(2,2,3);
101:    plot(Elementt(:,1),Elementt(:,2),'-');
102:    hold on;
103:    plot(Raxist(:,1),Raxist(:,2),':');
104:    if S1 ~= 0
105:       arrows(dirSx,0,angle,S1,0,0,1,Elementt);
106:       arrows(dirSx,0,pi+angle,0,0,0,0,0);
107:    end
108:    if S2 ~= 0
109:       arrows(dirSy,0,pi2+angle,0,0,0,0,0);
110:       arrows(dirSy,0,-pi2+angle,0,S2,0,2,0);
111:    end
112:    title('Principal Normal Stresses');
113:    axis('off'); hold off;
114:    set(gca,'Xlim',[Rlimits(1) Rlimits(2)]);
115:    set(gca,'Ylim',[Rlimits(3) Rlimits(4)]);
116:    drawnow;
117:
118: %........................
119: %...Max/min shear stresses
120: %........................
121: angle=Tau_max_ang*degrad;
122: S1=Sigma_avg; T=Tau_mxmn;
123:
124: %...Arrowhead direction
125: dirSx=sign(Sigma_avg); dirSy=sign(Sigma_avg);
126: dirTxy=sign(Tau_mxmn);
127:
128: %...Rotate by angle
129: trans=[cos(angle) -sin(angle);
130:        sin(angle)  cos(angle)];
131: temp=trans*Element'; Elementt=temp';
132: temp=trans*Raxis';   Raxist=temp';
133:
134: subplot(2,2,4);
```

```
135:     plot(Elementt(:,1),Elementt(:,2),'-');
136:     hold on;
137:     plot(Raxist(:,1),Raxist(:,2),':');
138:     arrows(dirSx,0,angle,S1,0,0,1,0);
139:     arrows(dirSx,0,pi+angle,0,0,0,0,0);
140:     arrows(dirSy,0,pi2+angle,0,0,0,0,0);
141:     arrows(dirSy,0,-pi2+angle,0,S1,0,2,0);
142:     if T ~= 0
143:       arrows(-dirTxy,-1,angle,0,0,0,0,0);
144:       arrows(-dirTxy,-1,pi+angle,0,0,T,3, ...
145:             Elementt);
146:       arrows(dirTxy,-1,pi2+angle,0,0,0,0,0);
147:       arrows(dirTxy,-1,-pi2+angle,0,0,0,0,0);
148:     end
149:     title('Maximum Shear Stresses');
150:     axis('off'); hold off;
151:     set(gca,'Xlim',[Rlimits(1) Rlimits(2)]);
152:     set(gca,'Ylim',[Rlimits(3) Rlimits(4)]);
153:     drawnow;
```

Function arrows

```
 1: function arrows(Adirection,Atype,Angle, ...
 2:    Sigma_1,Sigma_2,Tau,Label,Elementt)
 3: %
 4: % arrows(Adirection,Atype,Angle, ...
 5: %          Sigma_1,Sigma_2,Tau,Label,Elementt)
 6: % ~~~~~~~~~~~~~~~~~~~~~~~~~~~~~~~~~~~~~~~~~~~
 7: % This function draws the arrows for a stress
 8: % element box.  Normal stresses should use
 9: % full arrowheads and shear stresses should
10: % use half arrowheads.  All arrows are
11: % drawn from two reference arrows:
12: %    a) horizontal line (normals)
13: %    b) vertical line (shears)
14: % and rotated as specified to produced the
15: % complete set of arrows for an element.
16: %
17: % Adirection     - +1, arrowhead to right/up
18: %                 -1, arrowhead to left/down
19: % Atype         - 0, full arrowhead
20: %                  1, half arrowhead
```

5.1. Transformation of Plane Stress

```
21: % Angle          - angle to rotate arrow
22: %                  by in radians, measured
23: %                  from + x-axis CCW
24: % Sigma_1        - 1st sigma value for label
25: % Sigma_2        - 2nd sigma value for label
26: % Tau            - tau value for label
27: % Label          - 1, place Sigma_1 label
28: %                  2, place Sigma_2 label
29: %                  3, place Tau label
30: % Elementt       - coordinates of transformed
31: %                  element
32: %
33: % User m functions called:  none
34: %-----------------------------------------------
35:
36: %...Arrowhead size
37: Ahead_length=.3; Ahead_width=.075;
38:
39: %...Basic definition for arrows
40: %...  (x0,y0) - starting point of arrow
41: %...  Arrow_length - length of arrow
42: if Atype == 0
43:    %...Normal
44:    x0=1; y0=0; Arrow_length=1.0;
45: else
46:    %...Shear
47:    x0=1.2; y0=-.8; Arrow_length=1.6;
48: end
49:
50: %...Line part of arrow
51: endx=x0+Arrow_length; endy=y0+Arrow_length;
52:
53: %...Labels
54: if Label == 1
55:    %...Sigma_1 label
56:    xoffset=0.1; yoffset=0.1;
57:    string=num2str(abs(Sigma_1));
58:    strpart=[endx+xoffset y0+yoffset];
59: elseif Label == 2
60:    %...Sigma_2 label
61:    xoffset=0.1; yoffset=0.1;
62:    string=num2str(abs(Sigma_2));
63:    strpart=[endx+xoffset y0+yoffset];
64: elseif Label == 3
```

```
65:    %...Tau label
66:    string=num2str(abs(Tau));
67:    %strpart=[1.1 -1.3];
68:    [xmx,xmxi]=max(Elementt(:,1));
69:    yxmx=Elementt(xmxi,2);
70:    strpart=[xmx+0.1 yxmx];
71: end
72:
73: %...Create coordinates
74: if Atype == 0
75:    %.............
76:    %...Full arrow
77:    %.............
78:    linepart=[x0 y0;endx y0];
79:    if Adirection == 1
80:       %...Right
81:       arpart=[endx y0; ...
82:               endx-Ahead_length y0-Ahead_width; ...
83:               endx-Ahead_length y0+Ahead_width];
84:    else
85:       %...Left
86:       arpart=[x0 y0; ...
87:               x0+Ahead_length y0-Ahead_width; ...
88:               x0+Ahead_length y0+Ahead_width];
89:    end
90: else
91:    %.............
92:    %...Half arrow
93:    %.............
94:    linepart=[x0 y0;x0 endy];
95:    if Adirection == 1
96:       %...Up
97:       arpart=[x0 endy; ...
98:               x0+Ahead_width endy-Ahead_length; ...
99:               x0 endy-Ahead_length];
100:    else
101:       %...Down
102:       arpart=[x0 y0; ...
103:               x0+Ahead_width y0+Ahead_length; ...
104:               x0 y0+Ahead_length];
105:    end
106: end
107:
108: %...Generate transformation matrix
```

5.1. Transformation of Plane Stress

```
109: trans=[cos(Angle) -sin(Angle);
110:        sin(Angle)  cos(Angle)];
111:
112: %...Perform transformations
113: temp=trans*linepart'; linepart=temp;
114: temp=trans*arpart';  arpart=temp;
115:
116: fill(arpart(1,:),arpart(2,:),[1 1 1]);
117: if Label == 1 | Label == 2
118:    temp=trans*strpart'; strpart=temp;
119:    text(strpart(1),strpart(2),string);
120: elseif Label == 3
121:    text(strpart(1),strpart(2),string);
122: end
123: plot(linepart(1,:),linepart(2,:));
```

Function polarstr

```
 1: function polarstr(theta,rho,line_style)
 2: %
 3: % polarstr(theta,rho,line_style)
 4: % ~~~~~~~~~~~~~~~~~~~~~~~~~~~~~
 5: % This is a modified version of MATLAB's
 6: % standard plotting routine.  This version
 7: % has been tailored for this specific problem
 8: % and is not for general polar plotting.
 9: %
10: % theta      - the angle vector (radians)
11: % rho        - the radius vector
12: % line_style - one of MATLAB's linestyles
13: %
14: % User m functions called: none
15: %------------------------------------------
16:
17: % get hold state
18: hold_state = ishold;
19:
20: %...Only do grids if hold is off
21: if ~hold_state
22:    %...Make a radial grid
23:    hold on;
24:    hhh=plot([0 max(theta(:))], ...
```

```
25:            [0 max(abs(rho(:)))],'y:');
26:    v=[get(gca,'xlim') get(gca,'ylim')];
27:    ticks=length(get(gca,'ytick'));
28:    delete(hhh);
29:    %...Check radial limits and ticks
30:    rmin=0; rmax=v(4); rticks=ticks-1;
31:    if rticks > 5
32:      %...Can we reduce the number
33:      if rem(rticks,2) == 0
34:        rticks=rticks/2;
35:      elseif rem(rticks,3) == 0
36:        rticks=rticks/3;
37:      end
38:    end
39:    %...Define a circle
40:    th=0:pi/50:2*pi;
41:    xunit=cos(th); yunit=sin(th);
42:    %...Force points to lie on x/y axes
43:    inds=[1:(length(th)-1)/4:length(th)];
44:    xunits(inds(2:2:4))=zeros(2,1);
45:    yunits(inds(1:2:5))=zeros(3,1);
46:    rinc=(rmax-rmin)/rticks;
47:    for i=(rmin+rinc):rinc:rmax
48:      plot(xunit*i,yunit*i,'y:','linewidth',0.5);
49:      if (i ~= rmin+rinc) & (i ~= rmax)
50:        text(0,i+rinc/20,['  ' num2str(i)], ...
51:             'verticalalignment','bottom' );
52:      end
53:    end
54:    %...Plot spokes
55:    th=(1:6)*2*pi/12;
56:    cst=cos(th); snt=sin(th);
57:    cs=[-cst; cst]; sn=[-snt; snt];
58:    plot(rmax*cs,rmax*sn,'y:','linewidth',0.5);
59:    %...Annotate spokes in degrees
60:    rt=1.1*rmax;
61:    for i = 1:max(size(th))
62:      text(rt*cst(i),rt*snt(i), ...
63:           int2str(i*30), ...
64:           'horizontalalignment','center' );
65:      if i == max(size(th))
66:        loc=int2str(0);
67:      else
68:        loc=int2str(180+i*30);
```

5.2. Transformation of Plane Strain

```
69:     end
70:     text(-rt*cst(i),-rt*snt(i),loc, ...
71:          'horizontalalignment','center' );
72:   end
73:   %...Set axis limits
74:   axis(rmax*[-1 1 -1.1 1.1]);
75: end
76:
77: %...Transform data to Cartesian coordinates.
78: xx=rho.*cos(theta); yy=rho.*sin(theta);
79: plot(xx,yy,line_style);
```

5.2 Transformation of Plane Strain

The analysis of plane strain is analogous to the analysis of plane stress presented in the previous section. The difference is that the three values of stress (σ_x, σ_y, τ_{xy}) are replaced by three strain measures (ϵ_x, ϵ_y, γ_{xy}). Figure 5.7 indicates the sign convention which will be used for plane strain. This sign convention correlates with the sign convention for plane stress shown in Figure 5.1. The equations for the transformation of plane strain are identical to those for plane stress and can be duplicated by substituting ϵ_x for σ_x, ϵ_y for σ_y, and $\gamma_{xy}/2$ for τ_{xy}. In this section the equations will be summarized for plane strain and the development of a computer program is left as an exercise.

5.2.1 Rotation of Strain

Similar to the relations for plane stress, the strain measures are dependent on the set of axes chosen. Therefore, their values will change as the perpendicular reference axes are rotated. The equations for the strain values when rotated through an angle θ (θ is measured from the positive x-axis and is positive in the counterclockwise direction) are

$$\epsilon_\theta = \frac{\epsilon_x + \epsilon_y}{2} + \frac{\epsilon_x - \epsilon_y}{2}\cos(2\theta) - \frac{\gamma_{xy}}{2}\sin(2\theta)$$

$$\epsilon_{\theta+90°} = \frac{\epsilon_x + \epsilon_y}{2} - \frac{\epsilon_x - \epsilon_y}{2}\cos(2\theta) + \frac{\gamma_{xy}}{2}\sin(2\theta)$$

$$\gamma_\theta = (\epsilon_x - \epsilon_y)\sin(2\theta) + \gamma_{xy}\cos(2\theta)$$

The values for the maximum and minimum strain may be found from these relationships by differentiating ϵ with respect to θ and setting the resulting equation equal to zero. The two planes obtained from this equation are called the *principal axes of strain* and are located by

$$\tan(2\theta_p) = \frac{-\gamma_{xy}}{\epsilon_x - \epsilon_y}$$

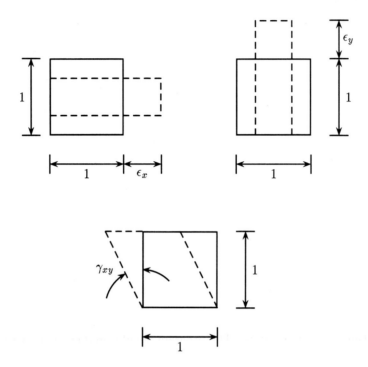

Figure 5.7. Sign Convention for Plane Strain

where the angle θ_p is measured from the positive x-axis to the principal planes. The above equation has two solutions for θ_p. One solution produces the angle to ϵ_{max} and the other solution produces the angle to ϵ_{min}. If the denominator of the above equation is zero then all angles of rotation for the element produce the same values for ϵ and are principal axes. The values for strain on the principal planes can be summarized as

$$\epsilon_{max} = \frac{\epsilon_x + \epsilon_y}{2} + \sqrt{\left(\frac{\epsilon_x - \epsilon_y}{2}\right)^2 + \left(\frac{\gamma_{xy}}{2}\right)^2}$$

$$\epsilon_{min} = \frac{\epsilon_x + \epsilon_y}{2} - \sqrt{\left(\frac{\epsilon_x - \epsilon_y}{2}\right)^2 + \left(\frac{\gamma_{xy}}{2}\right)^2}$$

where ϵ_{max} and ϵ_{min} are perpendicular and 90° apart and γ_{xy} on the principal planes is zero. The maximum and minimum value for γ_{xy} is produced on a plane 45° from

5.3. Strain Rosettes

the planes for ϵ_{max} and ϵ_{min}, or

$$(\gamma_{xy})_{max/min} = \pm 2\sqrt{\left(\frac{\epsilon_x - \epsilon_y}{2}\right)^2 + \left(\frac{\gamma_{xy}}{2}\right)^2}$$

and ϵ_x and ϵ_y on these planes is non-zero. The equation for the angle to the plane of maximum and minimum γ_{xy} is

$$\tan(2\alpha) = \frac{\epsilon_x - \epsilon_y}{\gamma_{xy}}$$

However, in practice this equation is rarely used since the $(\gamma_{xy})_{max}$ plane is $+45°$ from the plane of ϵ_{max} and the plane to $(\gamma_{xy})_{min}$ plane is $-45°$ from the plane of ϵ_{max}. Finally, if the denominator of the above equation is zero (i.e. $\gamma_{xy} = 0$) then the axes are principal strain axes.

5.2.2 Exercises

1. Develop a program, similar to **strglex**, which determines the strains on a prescribed plane. The program should produce plots which summarize the results.

2. Develop a program, similar to **prstrex**, which determines the principal strains and their associated angles. The program should produce plots which summarize the results.

3. Develop a program which plots the values for the three strain measures as they are rotated through 360°. The program should produce plots which summarize the results (similar to Figure 5.5).

4. Plot the quantity $\epsilon_x + \epsilon_y$ as it varies with θ. Can you make any conclusions from this plot?

5.3 Strain Rosettes

Strain gages [31, 36] are routinely used for measuring the strain in members under load. Figure 5.8 depicts the geometry for a typical strain gage. When the member is subjected to loading, the wire filaments of the gage either contract or stretch. The gages are supplied with an electrical current and their deformation induces a change in electrical resistance. These changes in electrical resistance can be converted directly to member strains with the appropriate instrumentation.

The direction of strain measurement may be easily recognized by recalling the equation for deflection in an axially loaded member, or

$$\Delta = \frac{PL}{AE}$$

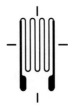

Figure 5.8. Strain Gage

where P is the applied load, L is the length of the member, A is the area of the cross section, and E is the modulus of elasticity for the material. Therefore, the direction of strain measurement corresponds to the direction of the long filaments in the gage.

A single strain gage only measures the deformation in a single direction. Frequently, the principal stresses and strains are desired for the loaded member. Multiple gages are employed to make the measurements required to determine the principal strains and stresses. This multigage configuration is called a *strain rosette*. Three types of rosettes are commonly used. Figure 5.9 depicts the layout for rectangular, delta, and T-delta rosettes.

5.3.1 Basic Mathematical Relationships

The fundamental relationship for transforming strain was presented earlier in this chapter, or

$$\epsilon_\theta = \frac{\epsilon_x + \epsilon_y}{2} + \frac{\epsilon_x - \epsilon_y}{2}\cos(2\theta) + \frac{\gamma_{xy}}{2}\sin(2\theta)$$

This leads to the following expressions for the principal strains.

$$\epsilon_{\max} = \frac{\epsilon_x + \epsilon_y}{2} + \frac{1}{2}\sqrt{(\epsilon_x - \epsilon_y)^2 + \gamma_{xy}^2}$$

$$\epsilon_{\min} = \frac{\epsilon_x + \epsilon_y}{2} - \frac{1}{2}\sqrt{(\epsilon_x - \epsilon_y)^2 + \gamma_{xy}^2}$$

$$\gamma_{\max/\min} = \pm\sqrt{(\epsilon_x - \epsilon_y)^2 + \gamma_{xy}^2}$$

$$\theta_p = \frac{1}{2}\tan^{-1}\left(\frac{\gamma_{xy}}{\epsilon_x - \epsilon_y}\right)$$

where θ_p is measured from the x-axis to the maximum principal strain. Positive angles are measured in a counterclockwise direction.

5.3. Strain Rosettes

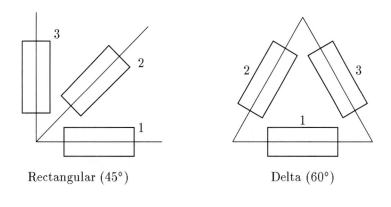

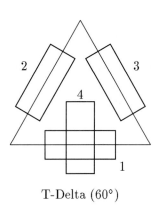

Figure 5.9. Strain Rosettes

5.3.2 The Rectangular Rosette

The rectangular rosette is formed by placing gages at $\theta_1 = 0°$, $\theta_2 = 45°$, and $\theta_3 = 90°$. Substituting these values into the equation for ϵ_θ yields

$$\epsilon_1 = \epsilon_x = \frac{\epsilon_x + \epsilon_y}{2} + \frac{\epsilon_x - \epsilon_y}{2}$$

$$\epsilon_2 = \frac{\epsilon_x + \epsilon_y}{2} + \frac{\gamma_{xy}}{2}$$

$$\epsilon_3 = \epsilon_y = \frac{\epsilon_x + \epsilon_y}{2} - \frac{\epsilon_x - \epsilon_y}{2}$$

which can be summarized as

$$\epsilon_x = \epsilon_1 \qquad \epsilon_y = \epsilon_3$$

$$\gamma_{xy} = 2\epsilon_2 - \epsilon_1 - \epsilon_3$$

The principal strains can be determined using these identities, or

$$\epsilon_{max} = \frac{\epsilon_1 + \epsilon_3}{2} + \frac{1}{2}\sqrt{(\epsilon_1 - \epsilon_3)^2 + (2\epsilon_2 - \epsilon_1 - \epsilon_3)^2}$$

$$\epsilon_{min} = \frac{\epsilon_1 + \epsilon_3}{2} - \frac{1}{2}\sqrt{(\epsilon_1 - \epsilon_3)^2 + (2\epsilon_2 - \epsilon_1 - \epsilon_3)^2}$$

$$\gamma_{max/min} = \pm \sqrt{(\epsilon_1 - \epsilon_3)^2 + (2\epsilon_2 - \epsilon_1 - \epsilon_3)^2}$$

$$\theta_p = \frac{1}{2} \tan^{-1}\left(\frac{2\epsilon_2 - \epsilon_1 - \epsilon_3}{\epsilon_1 - \epsilon_3}\right)$$

The corresponding values for the principal stresses are

$$\sigma_{max} = \frac{E(\epsilon_1 + \epsilon_3)}{2(1-\mu)} + R$$

$$\sigma_{min} = \frac{E(\epsilon_1 + \epsilon_3)}{2(1-\mu)} - R$$

$$\tau_{max/min} = \pm R$$

where

$$R = \frac{E}{2(1+\mu)}\sqrt{(\epsilon_1 - \epsilon_3)^2 + (2\epsilon_2 - \epsilon_1 - \epsilon_3)^2}$$

5.3.3 The Delta Rosette

The delta rosette is formed by placing gages at $\theta_1 = 0°$, $\theta_2 = 60°$, and $\theta_3 = 120°$. Substituting these values into the equation for ϵ_θ yields

$$\epsilon_1 = \epsilon_x = \frac{\epsilon_x + \epsilon_y}{2} + \frac{\epsilon_x - \epsilon_y}{2}$$

$$\epsilon_2 = \frac{\epsilon_x + \epsilon_y}{2} - \frac{\epsilon_x - \epsilon_y}{4} + \frac{\gamma_{xy}}{4}\sqrt{3}$$

$$\epsilon_3 = \frac{\epsilon_x + \epsilon_y}{2} - \frac{\epsilon_x - \epsilon_y}{4} - \frac{\gamma_{xy}}{4}\sqrt{3}$$

5.3. Strain Rosettes

which can be summarized as

$$\epsilon_x = \epsilon_1$$

$$\epsilon_y = \frac{-\epsilon_1 + 2\epsilon_2 + 2\epsilon_3}{3}$$

$$\gamma_{xy} = \frac{2(\epsilon_2 - \epsilon_3)}{\sqrt{3}}$$

The principal strains can be determined using these identities, or

$$\epsilon_{max} = \frac{\epsilon_1 + \epsilon_2 + \epsilon_3}{3} + \sqrt{\left(\epsilon_1 - \frac{\epsilon_1 + \epsilon_2 + \epsilon_3}{3}\right)^2 + \left(\frac{\epsilon_2 - \epsilon_3}{\sqrt{3}}\right)^2}$$

$$\epsilon_{min} = \frac{\epsilon_1 + \epsilon_2 + \epsilon_3}{3} - \sqrt{\left(\epsilon_1 - \frac{\epsilon_1 + \epsilon_2 + \epsilon_3}{3}\right)^2 + \left(\frac{\epsilon_2 - \epsilon_3}{\sqrt{3}}\right)^2}$$

$$\gamma_{max/min} = \pm 2 \sqrt{\left(\epsilon_1 - \frac{\epsilon_1 + \epsilon_2 + \epsilon_3}{3}\right)^2 + \left(\frac{\epsilon_2 - \epsilon_3}{\sqrt{3}}\right)^2}$$

$$\theta_p = \frac{1}{2} \tan^{-1} \left(\frac{\frac{1}{\sqrt{3}}(\epsilon_2 - \epsilon_3)}{\epsilon_1 - \frac{\epsilon_1 + \epsilon_2 + \epsilon_3}{3}} \right)$$

The corresponding values for the principal stresses are

$$\sigma_{max} = \frac{E(\epsilon_1 + \epsilon_2 + \epsilon_3)}{3(1 - \mu)} - R$$

$$\sigma_{min} = \frac{E(\epsilon_1 + \epsilon_2 + \epsilon_3)}{3(1 - \mu)} - R$$

$$\tau_{max/min} = \pm R$$

where

$$R = \frac{E}{1 + \mu} \sqrt{\left(\epsilon_1 - \frac{\epsilon_1 + \epsilon_2 + \epsilon_3}{3}\right)^2 + \left(\frac{\epsilon_2 - \epsilon_3}{\sqrt{3}}\right)^2}$$

5.3.4 The T-delta Rosette

The T-delta rosette is identical to the delta rosette ($\theta_1 = 0°$, $\theta_2 = 60°$, and $\theta_3 = 120°$) except that an extra gage is mounted perpendicular ($\theta_4 = 90°$) but on top of one of the other gages. The fourth gage can be used as a check gage or directly to measure the principal strains and stresses using the following equations.

$$\epsilon_{max} = \frac{\epsilon_1 + \epsilon_4}{2} + \frac{1}{2}\sqrt{(\epsilon_1 - \epsilon_4)^2 + \frac{4}{3}(\epsilon_2 - \epsilon_3)^2}$$

$$\epsilon_{min} = \frac{\epsilon_1 + \epsilon_4}{2} - \frac{1}{2}\sqrt{(\epsilon_1 - \epsilon_4)^2 + \frac{4}{3}(\epsilon_2 - \epsilon_3)^2}$$

$$\gamma_{max/min} = \pm\sqrt{(\epsilon_1 - \epsilon_4)^2 + \frac{4}{3}(\epsilon_2 - \epsilon_3)^2}$$

$$\theta_p = \frac{1}{2}\tan^{-1}\left(\frac{2(\epsilon_2 - \epsilon_3)}{\sqrt{3}(\epsilon_1 - \epsilon_4)}\right)$$

The corresponding values for the principal stresses are

$$\sigma_{max} = \frac{E}{2}\left[\frac{\epsilon_1 + \epsilon_4}{1 - \mu} + \frac{1}{1 + \mu}\sqrt{(\epsilon_1 - \epsilon_4)^2 + \frac{4}{3}(\epsilon_2 - \epsilon_3)^2}\right]$$

$$\sigma_{min} = \frac{E}{2}\left[\frac{\epsilon_1 + \epsilon_4}{1 - \mu} - \frac{1}{1 + \mu}\sqrt{(\epsilon_1 - \epsilon_4)^2 + \frac{4}{3}(\epsilon_2 - \epsilon_3)^2}\right]$$

$$\tau_{max/min} = \pm\frac{E}{2(1 + \mu)}\sqrt{(\epsilon_1 - \epsilon_4)^2 + \frac{4}{3}(\epsilon_2 - \epsilon_3)^2}$$

5.3.5 Compact Form of Rosette Relationships for Principal Stress

Table 5.2 [36] provides a set of constants which can be used with the associated equations below. The use of these constants greatly simplifies the calculation of the principal stress for the three types of rosettes.

$$\sigma_{max} = \frac{E}{1 - \mu}A + \frac{E}{1 + \mu}\sqrt{B^2 + C^2}$$

$$\sigma_{min} = \frac{E}{1 - \mu}A - \frac{E}{1 + \mu}\sqrt{B^2 + C^2}$$

5.3. Strain Rosettes

	Rectangular	delta	T-delta
A	$\frac{\epsilon_1+\epsilon_3}{2}$	$\frac{\epsilon_1+\epsilon_2+\epsilon_3}{3}$	$\frac{\epsilon_1+\epsilon_4}{2}$
B	$\frac{\epsilon_1-\epsilon_3}{2}$	$\epsilon_1 - \frac{\epsilon_1+\epsilon_2+\epsilon_3}{3}$	$\frac{\epsilon_1-\epsilon_4}{2}$
C	$\frac{2\epsilon_2-\epsilon_1-\epsilon_3}{2}$	$\frac{\epsilon_2-\epsilon_3}{\sqrt{3}}$	$\frac{\epsilon_2-\epsilon_3}{\sqrt{3}}$

Table 5.2. Constants for Rosette Stress Equations

$$\tau_{\text{max/min}} = \pm \frac{E}{1+\mu}\sqrt{B^2+C^2}$$

$$\theta_p = \frac{1}{2}\tan^{-1}\left(\frac{C}{B}\right)$$

5.3.6 Program to Determine Rosette Strains and Stresses

The program which calculates the principal stresses due to measurements obtained the three types of strain rosettes is straight forward. A summary of the program is provided in Table 5.3. Sample output is also provided in Section 5.3.8 for each of the three types of rosettes. Once again a computer program has been used to simplify the tedious task of performing repetitious calculations.

5.3.7 Exercises

1. Develop a function which produces a plot of the original rosette (handle all three types of rosettes). Overlay on this plot a plot of the exaggerated shape of the rosettes induced by the strains. A rectangle can be used to represent each individual strain gage in a rosette.

Table 5.3. Description of Code in Example **rosette**

Routine	Line	Operation
rosette		script file to execute program.
	22	select one of the example problems.
	23-35	define problem inputs.
	39-78	calculate principal strains and principal stress constants for specified rosette type.
	80-86	calculate principal stresses and associated angles.
	88-115	output the results.

5.3.8 Program Output and Code

Output from Example rosette

```
Rosette Analysis
----------------
Input:
  Modulus of elasticity: 3e+07
  Poisson's ratio:       0.3
  Rosette type:          1 (Rectangular)
  Strains:
    e1:                  0.000285
    e2:                  6.5e-05
    e3:                  0.000102

Principal stresses:
  Sigma-1:    11933.2
  Sigma-2:    4652.51
  Tau-12:     3640.34
  Sigma_ang:  -27.2734 degs

Principal strains:
  e-1:        0.000351248
  e-2:        3.57518e-05
  e-12:       0.000315496
```

5.3. Strain Rosettes

```
Rosette Analysis
----------------
Input:
  Modulus of elasticity: 3e+07
  Poisson's ratio:       0.3
  Rosette type:          2 (Delta)
  Strains:
    e1:                  0.000374
    e2:                  -0.000135
    e3:                  0.000227

Principal stresses:
  Sigma-1:    13637.5
  Sigma-2:    -323.252
  Tau-12:     6980.4
  Sigma_ang:  -21.8526 degs

Principal strains:
  e-1:        0.000457817
  e-2:        -0.00014715
  e-12:       0.000604968


Rosette Analysis
----------------
Input:
  Modulus of elasticity: 3e+07
  Poisson's ratio:       0.3
  Rosette type:          3 (T-Delta)
  Strains:
    e1:                  0.000225
    e2:                  0.000305
    e3:                  -0.000274
    e4:                  -6.5e-05

Principal stresses:
  Sigma-1:    11837.3
  Sigma-2:    -4980.17
  Tau-12:     8408.74
  Sigma_ang:  33.2754 degs
```

Principal strains:
  e-1:   0.000444379
  e-2:   -0.000284379
  e-12:   0.000728758

Script File rosette

```
1:  % Example: rosette
2:  % ~~~~~~~~~~~~~~~~
3:  % This program analyzes for the stresses or
4:  % strains in a plate using Hooke's law.
5:  %
6:  % Data is defined in the declaration statements
7:  % below, where:
8:  %
9:  % Emod      - modulus of elasticity
10: % mu        - Poisson's ratio
11: % Rosette   - =1, rectangular rosette
12: %             =2, delta rosette
13: %             =3, T-delta
14: % e1,e2,e3  - three strain measures
15: % e4        - required for T-delta rosettes
16: %
17: % User m functions required: none
18: %-----------------------------------------------
19:
20: clear;
21: %...Input definitions
22: Problem=1;
23: if Problem == 1
24:    %...Rectangular
25:    Rosette=1; Emod=30e6; mu=0.3;
26:    e1=285e-6; e2=65e-6; e3=102e-6;
27: elseif Problem == 2
28:    %...Delta
29:    Rosette=2; Emod=30e6; mu=0.3;
30:    e1=374e-6; e2=-135e-6; e3=227e-6;
31: elseif Problem == 3
32:    %...T-Delta
33:    Rosette=3; Emod=30e6; mu=0.3;
34:    e1=225e-6; e2=305e-6; e3=-274e-6; e4=-65e-6;
35: end
36:
```

5.3. Strain Rosettes

```
37: raddeg=180/pi;
38:
39: if Rosette == 1
40:    %............................
41:    %...Rectangular (3 @ 45 degs)
42:    %............................
43:    %...Strain calculations
44:    e_avg=(e1+e3)/2;
45:    Eradius=sqrt((e1-e3)^2+(2*e2-(e1+e3))^2);
46:    e_max=e_avg+0.5*Eradius;
47:    e_min=e_avg-0.5*Eradius; gamma_max=Eradius;
48:    %...Stress constants
49:    A=(e1+e3)/2; B=(e1-e3)/2;
50:    C=(2*e2-(e1+e3))/2;
51: elseif Rosette == 2
52:    %......................
53:    %...Delta (3 @ 60 degs)
54:    %......................
55:    %...Strain calculations
56:    e_avg=(e1+e2+e3)/3;
57:    Eradius=sqrt((e1-(e1+e2+e3)/3)^2+ ...
58:            ((e2-e3)/sqrt(3))^2);
59:    e_max=e_avg+Eradius; e_min=e_avg-Eradius;
60:    gamma_max=2*Eradius;
61:    %...Stress constants
62:    A=(e1+e2+e3)/3; B=e1-(e1+e2+e3)/3;
63:    C=(e2-e3)/sqrt(3);
64: elseif Rosette == 3
65:    %....................................
66:    %...T-delta (3 @ 60 degs, 1 @ 0 degs)
67:    %....................................
68:    %...Strain calculations
69:    e_avg=(e1+e4)/2;
70:    Eradius=sqrt((e1-e4)^2+4/3*(e2-e3)^2);
71:    e_max=e_avg+Eradius/2;
72:    e_min=e_avg-Eradius/2; gamma_max=Eradius;
73:    %...Stress constants
74:    A=(e1+e4)/2; B=(e1-e4)/2; C=(e2-e3)/sqrt(3);
75: else
76:    fprintf( ...
77:    '\n\nERROR: Undefined rosette type\n');
78: end
79:
80: %...Stress calculations
```

```
81: Savg=Emod/(1-mu)*A;
82: Sradius=Emod/(1+mu)*sqrt(B^2+C^2);
83: Sigma_max=Savg+Sradius;
84: Sigma_min=Savg-Sradius; Tau_max=Sradius;
85: Sigma_ang_mxmn=0.5*atan2(C,B);
86: Sigma_ang_mxmn=Sigma_ang_mxmn*raddeg;
87:
88: fprintf('\n\nRosette Analysis');
89: fprintf(  '\n----------------');
90: fprintf('\n\nInput:');
91: fprintf('\n  Modulus of elasticity: %g',Emod);
92: fprintf('\n  Poisson''s ratio:       %g',mu);
93: fprintf('\n  Rosette type:          %g ', ...
94:         Rosette);
95: if Rosette == 1, fprintf('(Rectangular)'); end
96: if Rosette == 2, fprintf('(Delta)'); end
97: if Rosette == 3, fprintf('(T-Delta)'); end
98: fprintf(  '\n  Strains:');
99: fprintf(  '\n     e1:                  %g ',e1);
100: fprintf(  '\n     e2:                  %g ',e2);
101: fprintf(  '\n     e3:                  %g ',e3);
102: if Rosette == 3
103:    fprintf('\n     e4:                  %g ',e4);
104: end
105: fprintf('\n\nPrincipal stresses:');
106: fprintf(  '\n  Sigma-1:    %g',Sigma_max);
107: fprintf(  '\n  Sigma-2:    %g',Sigma_min);
108: fprintf(  '\n  Tau-12:     %g',Tau_max);
109: fprintf(  '\n  Sigma_ang: %g',Sigma_ang_mxmn);
110: fprintf(  ' degs');
111: fprintf('\n\nPrincipal strains:');
112: fprintf(  '\n  e-1:        %g',e_max);
113: fprintf(  '\n  e-2:        %g',e_min);
114: fprintf(  '\n  e-12:       %g',gamma_max);
115: fprintf('\n');
```

Chapter 6

Stresses in Members of Polygonal Cross Section

6.1 Moment Angle Required to Induce Maximum Normal Stress

The flexure formula for a symmetrical cross section as shown in Figure 6.1a can be written as

$$\sigma(x,y) = \frac{M_x y}{I_x} - \frac{M_y x}{I_y}$$

where M_x and M_y are moments about their respective axes, I_x and I_y are the corresponding geometrical properties for the cross section relative to the centroidal axes, and (x,y) is the location on the cross section[1]. The moment vector acting on the cross section produces axial components defined by

$$M_x = M\cos\theta \qquad M_y = M\sin\theta$$

This flexure formula can be expanded for use with unsymmetrical cross sections (see Figure 6.1b) and has the form

$$\sigma(x,y) = \left(\frac{M_x I_y + M_y I_{xy}}{I_x I_y - I_{xy}^2}\right) y - \left(\frac{M_y I_x + M_x I_{xy}}{I_x I_y - I_{xy}^2}\right) x$$

which is known as the *generalized flexure formula*. Because the cross section is unsymmetrical the product of inertia, I_{xy}, appears in the generalized flexure formula. Frequently a cross section is subjected to a combination of bending and axial forces for which the resultant stress at any point (x,y) on the cross section can be determined using

$$\sigma(x,y) = \frac{N}{A} + \left(\frac{M_x I_y + M_y I_{xy}}{I_x I_y - I_{xy}^2}\right) y - \left(\frac{M_y I_x + M_x I_{xy}}{I_x I_y - I_{xy}^2}\right) x$$

where N is the axial load applied at the centroid of the cross section and A is the area of the cross section. This equation can be written in terms of the moment

---

[1] The negative sign denotes that a positive moment M_y causes a compressive stress.

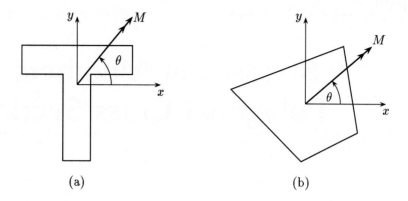

Figure 6.1. Cross Section Subjected to Moment Vector

vector M acting on the cross section and becomes

$$\sigma(x,y) = \frac{N}{A} + \left(\frac{M(\cos\theta)I_y + M(\sin\theta)I_{xy}}{I_x I_y - I_{xy}^2}\right) y - \left(\frac{M(\sin\theta)I_x + M(\cos\theta)I_{xy}}{I_x I_y - I_{xy}^2}\right) x$$

For a general polygonal cross section the stress with the greatest magnitude (either tensile or compressive) resulting from the moment vector will occur at one of the corner nodes. Since x and y are constant for a given corner, the angle for the moment vector which produces this extreme stress for any corner node can be found by differentiating the stress equation with respect to the angle θ and setting the result to zero, or

$$\frac{d\sigma}{d\theta} = 0 = \left(\frac{-M(\sin\theta)I_y + M(\cos\theta)I_{xy}}{I_x I_y - I_{xy}^2}\right) y - \left(\frac{M(\cos\theta)I_x + M(-\sin\theta)I_{xy}}{I_x I_y - I_{xy}^2}\right) x$$

Rewriting this equation in terms of the *sin* and *cos* terms produces[2]

$$0 = -c_1(\sin\theta) + c_2(\cos\theta)$$

---

[2] Note that the magnitudes of the moment vector and the axial load do not influence the angle of extreme stress.

6.1. Moment Angle Required to Induce Maximum Normal Stress

where
$$c_1 = \frac{yI_y - xI_{xy}}{I_xI_y - I_{xy}^2} \qquad c_2 = \frac{yI_{xy} - xI_x}{I_xI_y - I_{xy}^2}$$

This equation can be more succinctly expressed by

$$\tan\theta = \frac{c_2}{c_1}$$

This equation produces two angles for θ. One angle corresponds to the angle of maximum bending stress and the other to the angle of minimum bending stress due to the moment loading. The two corresponding values of stress are equal in magnitude, opposite in direction, and are 180° apart. The combined stress in the cross section is found by superposing the stress contributed by the axial load with the bending stresses acting at angle θ. These combined stresses represent the extreme tensile and compressive stresses on the specified corner of the cross section.

6.1.1 Program to Determine Angle Producing Maximum Normal Stress

The program to calculate the angle and corresponding stresses which produces the extreme normal stresses is summarized in Table 6.1. The output from an example problem is provided in Section 6.1.3. The graphical output from the example problem is shown in Figures 6.2 and 6.3. Figure 6.3 plots the angle producing the extreme stress for each corner node versus the corresponding values of maximum and minimum stress. The plot of moment angles verifies that the angle between the extreme maximum and minimum stresses differs by 180°. Additionally, the plot of stress values shows the symmetry of the stresses. If there is no axial force the symmetry will be about the zero stress axis, otherwise the symmetry is about a line shifted by the value of the axial stress. Figure 6.4 plots the variation of stress for each corner node for all angles of the symmetric angle example included with the program. (The creation of this plot is provided as an exercise.)

This program is dependent on the capability to calculate the geometrical properties. The functions necessary to calculate the geometrical properties of a polygon were presented in Chapter 4. Program **mxnorex** utilizes these routines to calculate the necessary geometrical properties.

Routine	Line	Operation
mxnorex		script file to execute program.
	24	select one of the example problems.
	25-44	define the input parameters.
	47	initialize some parameters.
	50	compute the geometrical properties.
	53-54	shift properties to centroidal axes.
	55	shift coordinates to centroidal axes.
	58-60	calculate some constants.
	69-70	calculate the bending stress.
	73-74	calculate the combined stress.
	76	identify the extreme case of bending stress.
	78-89	loop on each polygon corner and match the angles with the maximum and minimum stresses.
	90-91	convert angles to degrees.
	93-138	output the results.
	140-162	plot results.
prop		function which calculates the geometrical properties of a polygon. (See Chapter 4.)
shftprop		function to shift geometrical properties to centroidal axes. (See Chapter 4.)

Table 6.1. Description of Code in Example **mxnorex**

6.1.2 Exercises

1. Write a program which plots the variation of stress produced only by the moment vector as it varies with θ. Plot the curves for each corner node on the same graph. Are the maximum and minimum stresses equal in magnitude? Do they differ by 180°?

2. Write a program which plots the stress as it varies with θ as shown in Figure 6.4. Include the contributions from both the moment vector and the axial load. Are the maximum and minimum stresses equal in magnitude? Do they differ by 180°?

3. The neutral axis is defined as the axis of zero stress and its orientation determines the direction in which deflections occur (since the plane of bending is perpendicular to the neutral axis). For the angle of θ which produces the maximum tensile stress due to the combined loading, plot the line representing the neutral axis on the graph of the cross section using a dashed line.

6.1. Moment Angle Required to Induce Maximum Normal Stress

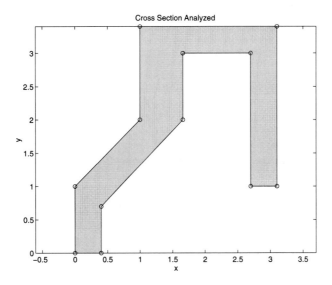

Figure 6.2. Cross Section Analyzed for Maximum Normal Stress

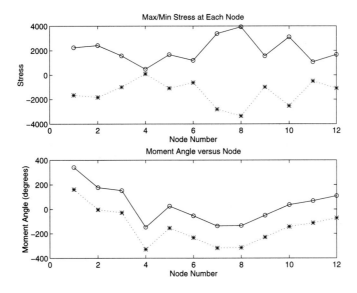

Figure 6.3. Maximum and Minimum Stress at Each Node

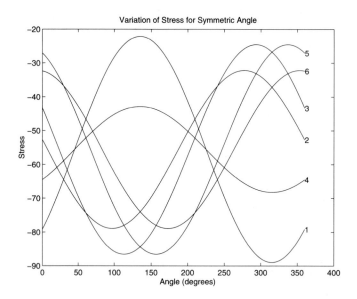

Figure 6.4. Variation of Stress for Symmetric Angle

Does the neutral axis always intersect with the cross section? Also plot the moment vector at angle θ using a dotted line.

4. Write a program which calculates the moment angle which produces the smallest value for extreme stress at any point in the cross section. HINT: Perform the same calculations as in Exercise 1 at one degree increments and select the angle which produces the smallest value of extreme stresses.

6.1.3 Program Output and Code

Output from Example mxnorex

```
     Angle for Moment Loading
Required to Produce Max/Min Stress
    in a Polygonal Cross Section
---------------------------------

Table of Nodes Forming Cross Section

  Node #          x             y
    1             0             0
    2            0.4            0
```

6.1. Moment Angle Required to Induce Maximum Normal Stress

3	0.4	0.7
4	1.66	2
5	1.66	3
6	2.7	3
7	2.7	1
8	3.1	1
9	3.1	3.4
10	1	3.4
11	1	2
12	0	1

Geometrical Properties
  A: 3.419
  Qx: 7.18777
  Qy: 5.58471
  Ix: 17.9966
  Iy: 12.3512
  Ixy: 13.3026
  x-bar: 1.63343
  y-bar: 2.1023

Geometrical Properties About Centroidal Axis
  Ix-bar: 2.8857
  Iy-bar: 3.2289
  Ixy-bar: 1.56189

Max/Min Stresses
  Axial force: 1000
  Applied moment: 3000

  Node 8 controls with:
    Maximum stress: 3933.03
      at an angle of: -134.496 degrees
    Minimum stress: -3348.06
      at an angle of: -314.496 degrees

Note: 1) Angles measured from positive x-axis,
         + for counter-clockwise
      2) + for tensile stress,
         - for compressive stress

Script File mxnorex

```
 1: % Example: mxnorex
 2: % ~~~~~~~~~~~~~~~~
 3: % This example determines the angles for
 4: % moment loading required to produce a
 5: % maximum or minimum stress in a polygonal
 6: % cross section.
 7: %
 8: % Data is defined in the declaration statements
 9: % below.
10: %
11: % x,y - vectors containing the coordinates for
12: %       the polygon nodes.  CCW fashion for
13: %       positive contributions, CW fashion
14: %       for negative contributions.
15: % P   - Applied axial force
16: % M   - Applied moment
17: %
18: % User m functions required:
19: %    prop, shftprop, genprint
20: %-------------------------------------------------
21:
22: clear;
23: %...Input definitions
24: Problem=1;
25: if Problem == 1
26:    %...Complex cross section
27:    x=[0 .4 .4 1.66 1.66 2.7 2.7 3.1 3.1 1 1 0];
28:    y=[0 0 .7 2 3 3 1 1 3.4 3.4 2 1];
29:    P=1000; M=3000;
30: elseif Problem == 2
31:    %...Symmetric angle
32:    x=[0 10 10 2 2 0]; y=[0 0 2 2 10 10];
33:    P=-2000; M=1000;
34: elseif Problem == 3
35:    %...Z section
36:    x=[-1 7 7 5 5 1 1 -7 -7 -5 -5 -1];
37:    y=[-10 -10 -6 -6 -8 -8 10 10 6 6 8 8];
38:    P=-2000; M=-2000;
39: else
40:    %...Channel
41:    x=[0 10 10 9 9 1 1 0];
```

6.1. Moment Angle Required to Induce Maximum Normal Stress   141

```
42:    y=[0 0 10 10 1 1 10 10];
43:    P=0; M=5000;
44: end
45:
46: %...Initialize
47: No_pts=length(x); raddeg=180/pi; Small=-1e20;
48:
49: %...Determine geometrical properties
50: [A,Qx,Qy,Ix,Iy,Ixy,xbar,ybar]=prop(x,y);
51:
52: %...Shift to centroidal axis
53: [Ixbar,Iybar,Ixybar]= ...
54:    shftprop(A,xbar,ybar,Ix,Iy,Ixy);
55: xs=x-xbar; ys=y-ybar;
56:
57: %...Constants for stress function
58: Denom=Ixbar*Iybar-Ixybar^2;
59: C1=Ixbar/Denom; C2=Iybar/Denom;
60: C3=Ixybar/Denom;
61:
62: %..................
63: %...Find worst case
64: %..................
65: Sigma_mxmn=Small;
66: K1=C2*ys-C3*xs; K2=C3*ys-C1*xs;
67: Beta=atan2(K2,K1);
68: cosine=cos(Beta); sine=sin(Beta);
69: Sigma=(C2*M*cosine+C3*M*sine).*ys- ...
70:       (C1*M*sine+C3*M*cosine).*xs;
71:
72: %...Total stress at each corner
73: Sigma_min=P/A-abs(Sigma);
74: Sigma_max=P/A+abs(Sigma);
75: %...Save the worst
76: [Sigma_mxmn,Index]=max(abs(Sigma));
77:
78: for i=1:No_pts
79:    %...Get the angles right
80:    if Sigma(i) < 0
81:       %...Minimum stress
82:       Angle_min(i)=Beta(i);
83:       Angle_max(i)=rem(Beta(i)+pi,2*pi);
84:    else
85:       %...Maximum stress
```

```
 86:       Angle_max(i)=Beta(i);
 87:       Angle_min(i)=Beta(i)-pi;
 88:    end
 89: end
 90: Angle_max=Angle_max*raddeg;
 91: Angle_min=Angle_min*raddeg;
 92:
 93: fprintf('\n\n     Angle for Moment Loading');
 94: fprintf('\nRequired to Produce Max/Min Stress');
 95: fprintf('\n   in a Polygonal Cross Section');
 96: fprintf('\n---------------------------------');
 97: fprintf( ...
 98:    '\n\nTable of Nodes Forming Cross Section');
 99: fprintf( ...
100:    '\n\n  Node #         x               y');
101: for i=1:No_pts
102:    fprintf('\n  %3.0f %12.5g  %12.5g', ...
103:            i,x(i),y(i));
104: end
105: fprintf('\n\nGeometrical Properties');
106: fprintf(  '\n  A:      %g',A);
107: fprintf(  '\n  Qx:     %g',Qx);
108: fprintf(  '\n  Qy:     %g',Qy);
109: fprintf(  '\n  Ix:     %g',Ix);
110: fprintf(  '\n  Iy:     %g',Iy);
111: fprintf(  '\n  Ixy:    %g',Ixy);
112: fprintf(  '\n  x-bar:  %g',xbar);
113: fprintf(  '\n  y-bar:  %g',ybar);
114: fprintf('\n\nGeometrical Properties About ');
115: fprintf(    'Centroidal Axis');
116: fprintf(  '\n  Ix-bar:  %g',Ixbar);
117: fprintf(  '\n  Iy-bar:  %g',Iybar);
118: fprintf(  '\n  Ixy-bar: %g',Ixybar);
119: fprintf('\n\nMax/Min Stresses');
120: fprintf('\n  Axial force:    %g',P);
121: fprintf('\n  Applied moment: %g',M);
122: fprintf('\n\n  Node %g controls with:',Index);
123: fprintf('\n    Maximum stress: %g', ...
124:          Sigma_max(Index));
125: fprintf( ...
126:    '\n      at an angle of: %g degrees', ...
127:    Angle_max(Index));
128: fprintf('\n    Minimum stress: %g', ...
129:          Sigma_min(Index));
```

6.2. Kern of a Compression Member

```
130:  fprintf( ...
131:     '\n       at an angle of: %g degrees', ...
132:     Angle_min(Index));
133:  fprintf('\n\nNote: 1) Angles measured from');
134:  fprintf(' positive x-axis,');
135:  fprintf(   '\n        + for counter-clockwise');
136:  fprintf('\n     2) + for tensile stress,');
137:  fprintf( ...
138:     '\n          - for compressive stress\n\n');
139:
140:  clf; x=[x(:);x(1)]; y=[y(:);y(1)];
141:  fill(x,y,'y'); hold on;
142:     plot(x,y,'-',x,y,'o');
143:     title('Cross Section Analyzed');
144:     xlabel('x'); ylabel('y');
145:     axis('equal'); hold off; drawnow;
146:  % genprint('mxnmsect');
147:     disp('Press key to continue'); pause
148:
149:  clf; x=1:No_pts;
150:  subplot(2,1,1);
151:     plot(x,Sigma_max,'-',x,Sigma_max,'o', ...
152:          x,Sigma_min,':',x,Sigma_min,'*');
153:     title('Max/Min Stress at Each Node')
154:     xlabel('Node Number');
155:     ylabel('Stress');
156:  subplot(2,1,2);
157:     plot(x,Angle_max,'-',x,Angle_max,'o', ...
158:          x,Angle_min,':',x,Angle_min,'*');
159:     title('Moment Angle versus Node')
160:     xlabel('Node Number');
161:     ylabel('Moment Angle (degrees)'); drawnow;
162:  % genprint('mxstress');
```

6.2 Kern of a Compression Member

When a compression member is loaded by an axial force applied through the centroid the entire cross section is uniformly stressed [3, 8]. A combination of axial compression and bending results when the member is subjected to an eccentrically applied loading as shown in Figure 6.5. Therefore, an eccentric load placed a suitable distance from the centroid will cause tension within the cross section. Many problems in structural mechanics require preventing the development of tensile forces. Examples include the design of prestressed concrete beams [6], the design of foot-

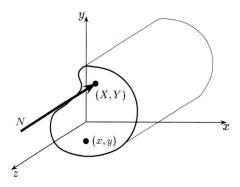

Figure 6.5. Axially Loaded Compression Member

ings [24], and the analysis and design of concrete dams [9]. This zone which defines the location where an axial load may be applied without inducing tensile stresses is called the kern[3] [15, 46]. In this section a program is presented which determines the kern for any geometry which can be represented as a n-sided polygon[4].

6.2.1 Mathematical Relationships

The flexure formula for a symmetrical cross section can be written as

$$\sigma(x, y) = \frac{M_x y}{I_x} - \frac{M_y x}{I_y}$$

where M_x and M_y are moments about their respective axes, I_x and I_y are the corresponding geometrical properties for the cross section relative to the centroidal axes, and (x, y) is the location on the cross section[5]. This relationship can be expanded for use with unsymmetrical cross sections and has the form

$$\sigma(x, y) = \left(\frac{M_x I_y + M_y I_{xy}}{I_x I_y - I_{xy}^2}\right) y - \left(\frac{M_y I_x + M_x I_{xy}}{I_x I_y - I_{xy}^2}\right) x$$

which is known as the *generalized flexure formula*. Because the cross section is unsymmetrical the product of inertia, I_{xy}, appears in the generalized flexure formula. Frequently a cross section is subjected to a combination of bending and axial

---

[3] The kern has also been referred to as the core [13, 32], a special case of the S-polygon [15, 26, 28], and the limit zone [17].

[4] Wilson and Turcotte [53] previously published a BASIC language implementation of this algorithm.

[5] The negative sign denotes that a positive moment M_y causes a compressive stress.

6.2. Kern of a Compression Member

forces for which the resultant stress at any point (x, y) on the cross section can be determined using

$$\sigma(x, y) = \frac{N}{A} + \left(\frac{M_x I_y + M_y I_{xy}}{I_x I_y - I_{xy}^2}\right) y - \left(\frac{M_y I_x + M_x I_{xy}}{I_x I_y - I_{xy}^2}\right) x$$

where N is the axial load and A is the area of the cross section. A special case of combined loading results when the cross section is subjected to a single eccentric axial load as shown in Figure 6.5. For this case the two moment vectors can be written as

$$M_x = NY \qquad M_y = -NX$$

where (X, Y) identifies the coordinates where the load acts and N is positive for tensile loads. Substituting these values into the general stress equation yields

$$\sigma(x, y) = \frac{N}{A} + N\left(\frac{YI_y - XI_{xy}}{I_x I_y - I_{xy}^2}\right) y - N\left(\frac{-XI_x + YI_{xy}}{I_x I_y - I_{xy}^2}\right) x$$

This equation is often expressed more compactly [46] as

$$\sigma(x, y) = \alpha + \beta x + \gamma y$$

where

$$\alpha = \frac{N}{A}$$

$$\beta = N\left(\frac{XI_x - YI_{xy}}{I_x I_y - I_{xy}^2}\right) x \qquad \gamma = N\left(\frac{YI_y - XI_{xy}}{I_x I_y - I_{xy}^2}\right) y$$

The three constants α, β, and γ depend on the loading quantities and the geometrical properties of the cross section. This equation can be rewritten in terms of the axial load, N, located at point (X, Y), or

$$\frac{\sigma}{N} = \tilde{\alpha} + \tilde{\beta} x + \tilde{\gamma} y$$

where

$$\tilde{\alpha} = \frac{1}{A} \qquad \tilde{\beta} = \frac{\beta}{N} \qquad \tilde{\gamma} = \frac{\gamma}{N}$$

The two constants $\tilde{\beta}$ and $\tilde{\gamma}$ are defined in terms of the geometrical properties and the location of the load, or

$$\tilde{\beta} = \frac{1}{I_x I_y - I_{xy}^2}(I_x X - I_{xy} Y)$$

$$\tilde{\gamma} = \frac{1}{I_x I_y - I_{xy}^2}(I_y Y - I_{xy} X)$$

It is evident from the generalized flexure formula that the normal stress on the cross section depends linearly on the coordinates of the point where stress is calculated and also depends linearly on the coordinates of the load placement. To determine the kern of the cross section it is necessary to calculate where loads can be placed which cause the stresses at the polygon corners to be zero. The condition of reciprocity [13] can be used quite effectively to determine the kern:

> If point $(\hat{x}_i, \hat{y}_i)$ represents one of the corner nodes of the polygon defining the cross section and the point of application of a load (see Figure 6.7) and the line P_i the corresponding neutral axis, then, if a load is applied anywhere on the line P_i, the neutral axis will pass through point $(\hat{x}_i, \hat{y}_i)$.

Therefore, by placing the load at any one of the polygon corners on the cross section the corresponding linear equation for the neutral axis can be found[6] by setting the stress equation to zero, or

$$0 = \tilde{\alpha} + \tilde{\beta} x + \tilde{\gamma} y$$

which can be rearranged in the more traditional form as

$$y = \left(-\frac{\tilde{\beta}}{\tilde{\gamma}}\right) x + \left(-\frac{\tilde{\alpha}}{\tilde{\gamma}}\right)$$

When the axial load is placed anywhere along the line defined by this equation (see Figure 6.6) the resulting stress on the corner node will be zero. This line defines one potential side of the kern polygon.

Figure 6.7 shows the boundary line produced by considering corner node $(\hat{x}_i, \hat{y}_i)$. This line, or plane, will be identified as the clipping plane P_i. The plane P_i divides the x-y plane into two parts. The side which contains corner $(\hat{x}_i, \hat{y}_i)$ is denoted as negative. Placing a load anywhere on the negative side of P_i will always produce compressive stress at $(\hat{x}_i, \hat{y}_i)$. Placing a load on the opposite side will cause tension. The kern of the cross section can be determined by considering all polygon corners and plotting the resulting region. Furthermore, axial loads placed inside this region will only induce compression in the cross section. The techniques required to accomplish this area identification entail polygon clipping [1, 10, 21, 42, 43] and are summarized in Appendix A.

6.2.2 Kern Construction

Assume that an arbitrary reference polygon (Typically, a large rectangle is used for the initial clipping polygon as shown in Figure 6.8.) is defined in terms of its corner coordinates. Also assume that a given clipping plane having a designated negative side intersects the polygon. Replacing the original reference polygon by the polygon remaining on the negative side of the clipping plane gives a so-called "clipped" polygon. As special cases, the clipped polygon could, of course, be identical with the original polygon or could be empty (zero area). Using this clipping mechanism, an algorithm to define the kern can be outlined as follows:

---

[6] Note that the kern is independent of the magnitude of the axial load N.

6.2. Kern of a Compression Member

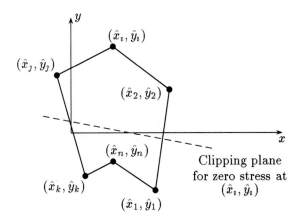

Figure 6.6. Cross Section of a General Polygon

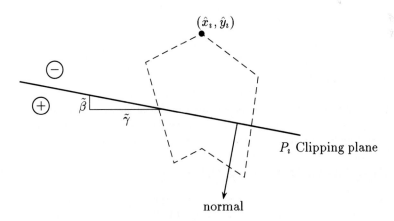

Figure 6.7. Clipping Plane

1. Locate the clipping plane P_1 corresponding to the first polygon corner $(\hat{x}_1, \hat{y}_1)$;
2. Define a large rectangle centered at the polygon centroid and clip this rectangle with respect to plane P_1;
3. Clip the polygon which is output from step 2 with respect to the plane P_2 associated with corner $(\hat{x}_2, \hat{y}_2)$;
4. Repeat the clipping process until all corners are considered.
5. The kern is the convex polygon remaining after all clipping is completed.

Figure 6.8 demonstrates the sequential clipping process for a simple geometry. In this figure the polygon drawn using solid lines is the original cross section and the polygon depicted by the dotted lines is the clipping polygon. A dashed line is used to indicate the clipping plane P_i. After the last corner node is processed the kern is identified and represented by the shaded polygon.

Appendix A.4 includes the necessary code to clip a reference polygon and produce the resulting "clipped" polygon. The clipping function requires the user to specify the clipping plane as the equation of a line in normal form[7] such that the corresponding normal vector is directed away from the clipped polygon as depicted in Figure 6.7. Recalling that the equation of the zero stress line P_i was

$$0 = \tilde{\alpha} + \tilde{\beta}x + \tilde{\gamma}y \quad \Longrightarrow \quad y = \left(-\frac{\tilde{\beta}}{\tilde{\gamma}}\right) x + \left(-\frac{\tilde{\alpha}}{\tilde{\gamma}}\right)$$

then the unit vector normal to this line has components

$$(\frac{\tilde{\beta}}{\Omega}, \frac{\tilde{\gamma}}{\Omega}, 0)$$

with

$$\Omega = \sqrt{\tilde{\beta}^2 + \tilde{\gamma}^2}$$

The clipping routine also requires that the normal line be directed away from the "clipped" zone (i.e. pointing towards the part of the reference polygon which will be clipped off). A simple method to insure this requirement is to take the dot product of the normal vector with a vector pointing at the corner node under consideration. If the dot product is positive the normal vector has the wrong direction.

6.2.3 Program to Determine the Kern

The program developed to find the kern of any polygonal cross section is summarized in Table 6.2. The output from an example problem (a structural angle) is provided in Section 6.2.5. The program includes several other example cross sections which demonstrate the flexibility offered by the technique for geometries with different construction characteristics, including

---

[7] Appendix A includes a discussion of the normal form of a line.

6.2. Kern of a Compression Member

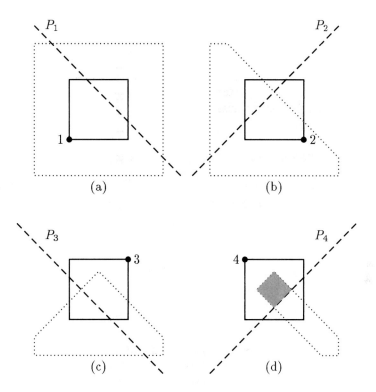

Figure 6.8. Clipping Sequence for Square Cross Section

- a standard channel where the kern is not within the cross section,
- a cross section with a cutout (or hole),
- a cross section comprised of two physically unconnected areas,
- and a cross section having a curved boundary (a three quarter circle).

The program to determine the kern of a polygonal cross section is dependent on the capability to calculate the geometrical properties. The algorithms necessary to calculate the geometrical properties of a polygon were presented in Chapter 4. Program **kern** utilizes these routines to calculate the necessary geometrical properties.

Figure 6.9 shows the graphical output generated by program **kern**. Both the original cross section and the kern are drawn. The kern polygon has been filled and each of the boundary points is indicated.

Routine	Line	Operation
kern		script file to execute program.
	22	select one of the example problems.
	23-46	define the input parameters.
	49	define a value for determining zero.
	52	compute the geometrical properties.
	55-56	shift properties to centroidal axes.
	57	shift coordinates to centroidal axes.
	60	define reference polygon for clipping.
	67-115	loop on each polygon corner node.
	68-91	define the unit normal.
	97-98	insure normal has correct direction.
	101-103	clip the reference polygon.
	106-114	update the reference polygon.
	120-151	output results.
	154-162	plot results.
prop		function which calculates the geometrical properties of a polygon. (See Chapter 4.)
shftprop		function to shift geometrical properties to centroidal axes. (See Chapter 4.)
clip		function to clip a polygon. (See Appendix A.)
refpoly		function which defines a large rectangle to use to clip against. (See Appendix A.)
inout		function to determine whether a point is above or below the clipping plane. (See Appendix A.)
intrsect		function which determines the point of intersection of two lines. (See Appendix A.)
circle		function which generates a polygonal representation for a circle. (See Appendix A.)

Table 6.2. Description of Code in Example **kern**

6.2.4 Exercises

1. The points where the principal axes of inertia intersect the boundaries of the kern are called the "kernel" points [13]. Add the capability to program **kern** to find the kernel points. Enhance the graphical output to include the principal inertia axes as dashed lines.

2. The kern for any cross section is independent of any particular axial loading. However, in many cases the requirement may be to determine if a load placed

6.2. Kern of a Compression Member

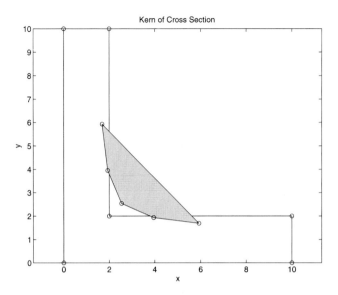

Figure 6.9. Kern of a Cross Section

at a specified location is within the kern. Develop a program which will automatically determine if an axial load with a specified coordinate location falls within the kern. HINT: The kern always produces a convex polygon. From the loading location extend a horizontal line either to the left or right. Count the number of times the kern boundary is crossed. An odd count value means the point is within the kern and an even count means the point is outside the kern. (Be careful to handle the case when the line intersects a corner point of the kern boundary.)

3. The use of the normal form for defining a line is important in the clipping routines used in this problem. Add the capability to program **kern** to create a plot containing the following for each iteration of the clipping loop: (a) draw the outline of the cross section using solid lines – place an "o" symbol on the corner node being considered in this iteration, (b) draw the outline of the clipping polygon using dotted lines, (c) draw the clipping line, and (d) draw the properly directed line normal to the clipping line – use an asterisk to indicate the end of the normal vector with the arrowhead. (This should be a set of figures similar to Figure 6.8 which also include the normal vectors.)

6.2.5 Program Output and Code

Output from Example kern

```
Kern of a Polygonal Cross Section
---------------------------------

Table of Nodes Forming Cross Section

    Node #          x               y
      1             0               0
      2             10              0
      3             10              2
      4             2               2
      5             2               10
      6             0               10

Geometrical Properties
  A:     36
  Qx:    116
  Qy:    116
  Ix:    688
  Iy:    688
  Ixy:   196
  x-bar: 3.22222
  y-bar: 3.22222

Geometrical Properties About Centroidal Axis
  Ix-bar:   314.222
  Iy-bar:   314.222
  Ixy-bar: -177.778

Table of Kern Boundary Nodes

   Node #         x               y
     1          5.931           1.6897
     2          1.6897          5.931
     3          1.9344          3.9508
     4          2.54            2.54
     5          3.9508          1.9344
```

6.2. Kern of a Compression Member

Script File kern

```
 1: % Example: kern
 2: % ~~~~~~~~~~~~~
 3: % This program determines the kern for an
 4: % axially loaded compression member described
 5: % by a polygonal cross section.
 6: %
 7: % Data is defined in the declaration statements
 8: % below, where:
 9: %
10: % x,y - vectors containing the coordinates for
11: %       the polygon nodes.  CCW fashion for
12: %       positive contributions, CW fashion
13: %       for negative contributions.
14: %
15: % User m functions required:
16: %     prop, shftprop, inout, intrsect, clip,
17: %     refpoly, circle, genprint
18: %-------------------------------------------------
19:
20: clear;
21: %...Input definitions
22: Problem=1;
23: if Problem == 1
24:   %...Symmetric angle
25:   x=[0 10 10 2 2 0]; y=[0 0 2 2 10 10];
26: elseif Problem == 2
27:   %...Z section
28:   x=[-1 7 7 5 5 1 1 -7 -7 -5 -5 -1];
29:   y=[-10 -10 -6 -6 -8 -8 10 10 6 6 8 8];
30: elseif Problem == 3
31:   %...Square with triangle cutout
32:   x=[0 10 10 0 0 5 2 8 5];
33:   y=[0 0 10 10 0 2 8 8 2];
34: elseif Problem == 4
35:   %...Two disconnected triangles
36:   x=[5 10 10 5 -10 -5 -10 -10];
37:   y=[0 -6 6 0 -6 0 6 -6];
38: elseif Problem == 5
39:   %...Channel
40:   x=[0 10 10 9 9 1 1 0];
41:   y=[0 0 10 10 1 1 10 10];
```

```
42:  else
43:    %...3/4 circular section
44:    [x,y]=circle(36,0,0,10);
45:    x=x(1:28); x(29)=0; y=y(1:28); y(29)=0;
46:  end
47:
48:  %...Initialize
49:  Epsilon=1e-8; NoPts=length(x);
50:
51:  %...Determine geometrical properties
52:  [A,Qx,Qy,Ix,Iy,Ixy,xbar,ybar]=prop(x,y);
53:
54:  %...Shift to centroidal axis
55:  [Ixbar,Iybar,Ixybar]= ...
56:    shftprop(A,xbar,ybar,Ix,Iy,Ixy);
57:  xs=x-xbar; ys=y-ybar;
58:
59:  %...Define polygon to clip against
60:  [NoClipPts,Xclip,Yclip]=refpoly(xs,ys);
61:
62:  %.........................................
63:  %...Loop on each corner of user geometry
64:  %...and clip reference polygon
65:  %.........................................
66:  Determinant=Ixbar*Iybar-Ixybar^2; alpha=1/A;
67:  for i=1:NoPts
68:    beta=(Ixbar*xs(i)-Ixybar*ys(i))/Determinant;
69:    gamma=(Iybar*ys(i)-Ixybar*xs(i))/Determinant;
70:    sfact=sign(gamma);
71:    if sfact == 0, sfact=1; end
72:    d=sqrt(beta^2+gamma^2)*sfact;
73:
74:    %...Note: The neutral axis for any candidate
75:    %...       point has the form:
76:    %...
77:    %...       alpha + beta*x + gamma*y = 0
78:    %...
79:    %...       The normal form is:
80:    %...
81:    %...       alpha    beta*x    gamma*y
82:    %...       ----- + ------ + ------- = 0
83:    %...         d        d        d
84:    %...
85:    %...       where:
```

6.2. Kern of a Compression Member

```
86:     %...
87:     %...         d=SIGN(beta)*SQRT(beta^2+gamma^2)
88:
89:     Nform(1)=alpha/d;   % perpendicular distance
90:     Nform(2)=beta/d;    % run of normal vector
91:     Nform(3)=gamma/d;   % rise of normal vector
92:
93:     %...Make sure line is defined such that
94:     %... a perpendicular (the normal vector for
95:     %... the clipping line) is directed away from
96:     %... clipping zone
97:     DotProduct=xs(i)*Nform(2)+ys(i)*Nform(3);
98:     if DotProduct >= 0, Nform=-Nform; end
99:
100:    %...Clip the reference polygon
101:    [NoNewPts,IsetClipPoly,NoInterpPts, ...
102:       Xclip,Yclip]=clip(Xclip,Yclip,NoClipPts, ...
103:       Nform,Epsilon);
104:
105:    %...Clip complete, reorder arrays
106:    Loop=NoClipPts+NoInterpPts;
107:    xtmp(1:Loop)=Xclip(1:Loop);
108:    ytmp(1:Loop)=Yclip(1:Loop);
109:    clear Xclip Yclip;
110:    for j=1:NoNewPts
111:       itmp=IsetClipPoly(j);
112:       Xclip(j)=xtmp(itmp); Yclip(j)=ytmp(itmp);
113:    end
114:    NoClipPts=NoNewPts;
115: end
116:
117: %...Convert back to original system
118: Xclip=Xclip+xbar; Yclip=Yclip+ybar;
119:
120: fprintf( ...
121:    '\n\nKern of a Polygonal Cross Section');
122: fprintf('\n--------------------------------');
123: fprintf( ...
124:    '\n\nTable of Nodes Forming Cross Section');
125: fprintf( ...
126:    '\n\n   Node #            x               y');
127: for i=1:NoPts
128:    fprintf('\n  %3.0f   %12.5g   %12.5g', ...
129:            i,x(i),y(i));
```

```
130:  end
131:  fprintf('\n\nGeometrical Properties');
132:  fprintf(  '\n   A:      %g',A);
133:  fprintf(  '\n   Qx:     %g',Qx);
134:  fprintf(  '\n   Qy:     %g',Qy);
135:  fprintf(  '\n   Ix:     %g',Ix);
136:  fprintf(  '\n   Iy:     %g',Iy);
137:  fprintf(  '\n   Ixy:    %g',Ixy);
138:  fprintf(  '\n   x-bar: %g',xbar);
139:  fprintf(  '\n   y-bar: %g',ybar);
140:  fprintf('\n\nGeometrical Properties About ');
141:  fprintf(    'Centroidal Axis');
142:  fprintf(  '\n   Ix-bar:  %g',Ixbar);
143:  fprintf(  '\n   Iy-bar:  %g',Iybar);
144:  fprintf(  '\n   Ixy-bar: %g',Ixybar);
145:  fprintf('\n\nTable of Kern Boundary Nodes');
146:  fprintf('\n\n  Node #           x                y');
147:  for i=1:NoClipPts
148:     fprintf('\n  %3.0f  %12.5g  %12.5g', ...
149:              i,Xclip(i),Yclip(i));
150:  end
151:  fprintf('\n');
152:
153:  %...Close end of polygon for plotting
154:  xp=[x(:);x(1)]; yp=[y(:);y(1)];
155:  clf;
156:  plot(xp,yp,'-',xp,yp,'o'); hold on;
157:     fill(Xclip,Yclip,'y');
158:     plot(Xclip,Yclip,'o',Xclip,Yclip,'-');
159:     title('Kern of Cross Section');
160:     xlabel('x'); ylabel('y');
161:     axis('equal'); hold off; drawnow;
162:  % genprint('kern');
```

6.3 Distribution of Shear Stresses Due to Bending

The shear stress which results from a shear load, V, over a cross section of a beam is described by the *shear formula*, or

$$\tau = \frac{VQ}{It}$$

where I is the moment of inertia for the cross section, Q is the first moment about the neutral axis of the area either above or below the point at which τ is determined, and t is the width of the cross section at the location of the stress. An additional

6.3. Distribution of Shear Stresses Due to Bending   157

quantity, called the *shear flow*, is commonly determined when analyzing the shear stresses due to bending and is defined by

$$q = \frac{VQ}{I}$$

and is valid for cross sections where the shear load acts in the direction of a principal axis of inertia.

Two programs are presented in this section which can be used to investigate the shear stresses which result from bending. The first program calculates the shear flow for a horizontal or vertical slice across the cross section. The second program graphs the distribution of shear stresses over the depth of a cross section using an automated method based on the polygon clipping technique described in Appendix A and used in the previous section of this chapter.

Both of the programs require the calculation of the geometrical properties of a polygonal cross section. The functions necessary to calculate the geometrical properties of a polygon were presented in Chapter 4.

6.3.1 Program to Determine the Shear Flow in a Cross Section

Figure 6.10 represents a cross section constructed using several parts. The cross section is symmetric about the vertical axis and therefore the shear flow can be calculated using the equation

$$q = \frac{VQ}{I}$$

The moment of inertia, I, for the cross section can be obtained using the methods developed in Chapter 4 and the shear, V, acting on the cross section is prescribed. Only the value for the first moment, Q, is required to determine the shear flow for any slice of the cross section.

As an example, consider a slice across the cross section an infinitesimal distance below where parts B and C are connected. The value of Q is simply the first moment of the parts C, D, and E about the neutral axis. Since parts C, D, and E combine to form a single polygon, the methods developed in Chapter 4 are employed to calculate Q. The program **shrstr1** and its associated subprograms are summarized in Table 6.3. Figure 6.11 shows the plot generated by the program. The shaded part of the cross section represents the area for which Q was determined. Section 6.3.4 includes the output for this example.

6.3.2 Program to Plot the Distribution of Shear Stresses

Determining the distribution of shear stresses over the depth of a cross section is important in beam design. A general method for performing this analysis on any cross section is beyond the scope of this book. However, an automated method can be implemented for a restricted class of cross sections. The properties of these beams are that

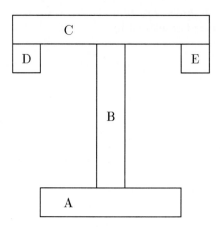

Figure 6.10. Cross Section Constructed Using Multiple Parts

- the cross section must be symmetric about the vertical axis, and
- the cross section, when sliced, must yield only two fully connected pieces (lower and upper). A cross section which violates this characteristic was presented in Figure 6.10. A slice performed at a plane connecting parts B and C would produce one upper piece and three lower pieces (D, E, and AB).

The shear stress at any elevation of the cross section is represented by

$$\tau = \frac{VQ}{It}$$

The shear force, V, is prescribed and the moment of inertia, I, can be easily determined for a polygonal cross section. The values for Q and t will depend on the elevation at which the shear stress is to be calculated. To plot the distribution of shear stresses over the depth of the beam an appropriate number of elevations must be analyzed. This can be performed automatically by performing an orderly sequence of horizontal slices, or "clips", over the depth of the beam and calculating the analogous shear stress for each slice. This requires an algorithm which can be used to "clip" a polygon. The clipping algorithm must produce the area required for the calculation of Q and the thickness T of the slice. Appendix A discusses the technique of polygon clipping and this method was used in the previous section concerning the calulation of the kern of a cross section.

The program **shrstr2** was developed to calculate and plot the distribution of shear stress in a cross section. This program and its associated subprograms are summarized in Table 6.3. Figure 6.12 shows the plot generated by the program. and the output for this example is included in Section 6.3.4.

6.3. Distribution of Shear Stresses Due to Bending   159

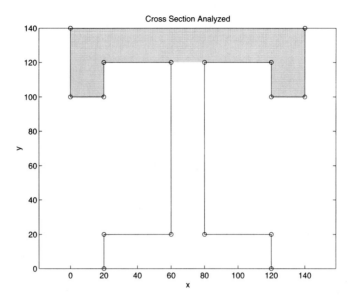

Figure 6.11. Shear Flow at an Interface

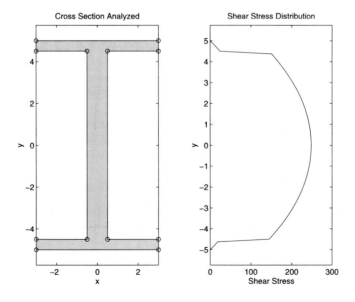

Figure 6.12. Distribution of Shear Stresses in a Cross Section

Routine	Line	Operation
shrstr1		script file to execute program.
	25	select one of the example problems.
	26-53	define the input parameters.
	56	compute the geometrical properties.
	57-59	shift properties to centroidal axes.
	60	shift coordinates to centroidal axes.
	63	compute the first moment of area.
	66	calculate the shear flow.
	68-103	output the results.
	105-106	close the polygons to plot.
	107-113	plot results.
shrstr2		script file to execute program.
	33	select one of the example problems.
	34-49	define the input parameters.
	51-54	initialize some parameters.
	57	compute the geometrical properties.
	59-60	shift properties to centroidal axes.
	61	shift coordinates to centroidal axes.
	63	find extreme y-axis coordinates.
	66	define the horizontal planes at which stress will be calculated.
	69	initialize loop parameters.
	70-79	loop on each horizontal plane.
	72-73	clip the original cross section and determine the polygon which defines the new "clipped" area.
	75-76	compute the first moment of this area.
	78	calculate the thickness of the area at the horizontal clipping plane.
	81	calculate the shear flow.
	82	calculate the shear stress.
	87-121	output the results.

continued on next page

6.3. Distribution of Shear Stresses Due to Bending

continued from previous page

Routine	Line	Operation
	123	close the polygon to plot.
	124-139	plot results.
genarea		function which determines the area which results from clipping the original cross section at a horizontal plane.
	32	define the numerical value of zero.
	45-47	define the coefficients for the normal form of clipping plane.
	49-50	use the original cross section to clip against.
	53-55	clip the cross section at the prescribed horizontal plane.
	58-65	create the set of nodes defining the clipped area by processing the results from the clipping function.
findt		function which calculates the thickness at the horizontal clipping plane.
	22	define the numerical value of zero.
	23	initialize loop parameters.
	24-32	loop on each node of polygon.
	29	check to see if two adjacent nodes have same y-coordinate.
	30	calculate the thickness.
prop		function which calculates the geometrical properties of a polygon. (See Chapter 4.)
shftprop		function to shift geometrical properties to centroidal axes. (See Chapter 4.)
clip		function to clip a polygon. (See Appendix A.)
inout		function to determine whether a point is above or below the clipping plane. (See Appendix A.)
intrsect		function which determines the point of intersection of two lines. (See Appendix A.)

Table 6.3. Description of Code in Examples **shrstr1** and **shrstr2**

6.3.3 Exercises

1. The average shear stress in a cross section can be defined by

$$\tau_{avg} = \frac{V}{A}$$

where A is the total area of the cross section. The maximum stress in the cross section is subsequently defined as

$$\tau_{max} = k_s \tau_{avg}$$

Write a function which calculates k_s for any polygonal cross section which is symmetric about the vertical axis.

6.3.4 Program Output and Code

Output from Example shrstr1

```
Shear Flow at Interface
----------------------

Table of Nodes Forming Cross Section

    Node #          x              y
       1           20              0
       2          120              0
       3          120             20
       4           80             20
       5           80            120
       6          120            120
       7          120            100
       8          140            100
       9          140            140
      10            0            140
      11            0            100
      12           20            100
      13           20            120
      14           60            120
      15           60             20
      16           20             20

Geometrical Properties
   A:     7600
   Qx:    612000
   Qy:    532000
   Ix:    6.88533e+007
```

6.3. Distribution of Shear Stresses Due to Bending

```
 Iy:     4.64533e+007
 Ixy:    4.284e+007
 x-bar:  70
 y-bar:  80.5263

Geometrical Properties About Centroidal Axis
 Ix-bar:   1.95712e+007
 Iy-bar:   9.21333e+006
 Ixy-bar:  0
```

Table of Nodes Forming Subarea

Node #	x	y
1	120	120
2	120	100
3	140	100
4	140	140
5	0	140
6	0	100
7	20	100
8	20	120

```
Shear (V):         1.5
First moment (Q):  162105
Shear flow (q):    0.0124243
```

Output from Example shrstr2

Shear Stresses in Cross Section

Table of Nodes Forming Cross Section

Node #	x	y
1	3	-5
2	3	-4.5
3	0.5	-4.5
4	0.5	4.5
5	3	4.5
6	3	5
7	-3	5
8	-3	4.5
9	-0.5	4.5
10	-0.5	-4.5

```
   11            -3           -4.5
   12            -3           -5
```

Geometrical Properties
   A: 15
   Qx: 0
   Qy: 0
   Ix: 196.25
   Iy: 18.75
   Ixy: 0
   x-bar: 0
   y-bar: 0

Geometrical Properties About Centroidal Axis
   Ix-bar: 196.25
   Iy-bar: 18.75
   Ixy-bar: 0

Shear (V): 2000

Clipping plane number 1
   y: -5
   First moment (Q):   0
   Thickness: 0
   Shear flow (q):   0
   Shear stress: 0

Clipping plane number 2
   y: -4.875
   First moment (Q):   3.70312
   Thickness: 6
   Shear flow (q):   37.7389
   Shear stress: 6.28981

   .
   . Material removed
   .

Clipping plane number 80
   y: 4.875
   First moment (Q):   3.70312
   Thickness: 6
   Shear flow (q):   37.7389
   Shear stress: 6.28981

6.3. Distribution of Shear Stresses Due to Bending   165

```
Clipping plane number 81
  y:                      5
  First moment (Q):       0
  Thickness:              0
  Shear flow (q):         0
  Shear stress:           0
```

Script File shrstr1

```
 1: % Example: shrstr1
 2: % ~~~~~~~~~~~~~~~~
 3: % This example determines the shear flow
 4: % on a subarea of a cross section.
 5: %
 6: % Data is defined in the declaration statements
 7: % below, where:
 8: %
 9: %-----------------------------------------------
10: % x,y   - vectors containing the coordinates
11: %           for the nodes of the cross section.
12: %           CCW fashion for positive
13: %           contributions, CW fashion for
14: %           negative contributions.
15: % xx,yy - vectors containing the coordinates
16: %           for the nodes of the subarea.
17: % V     - shear on the cross section
18: %
19: % User m functions required:
20: %     prop, shftprop, genprint
21: %-----------------------------------------------
22:
23: clear;
24: %...Input definitions
25: Problem=3;
26: if Problem == 1
27:    %...I-beam with different flanges
28:    x=[20 80 80 60  60 100 100   0   0  40 40 20];
29:    y=[ 0  0 20 20 100 100 120 120 100 100 20 20];
30:    V=1500;
31:    %...bottom flange
32:    xx=[20 80 80 20]; yy=[ 0  0 20 20];
33: elseif Problem == 2
```

```
34:     %...Box beam nailed together
35:     x=[0 120 120   0 0 20  20 100 100 20];
36:     y=[0   0 120 120 0 20 100 100  20 20];
37:     V=1200;
38:     %...top member
39:     xx=[20  100 100  20]; yy=[100 100 120 120];
40: else
41:     %...T-shaped I-beam
42:     %    1   2   3   4    5    6   7    8 ...
43:     %    9  10  11  12   13   14  15   16
44:     x=[20 120 120  80   80  120 120 140 ...
45:        140   0   0  20   20   60  60  20];
46:     y=[ 0   0  20  20  120  120 100 100 ...
47:        140 140 100 100  120  120  20  20];
48:     V=1.5;
49:     %...top section interface with web
50:     %       6   7   8    9  10  11  12  13
51:     xx=[120 120 140 140   0   0  20  20];
52:     yy=[120 100 100 140 140 100 100 120];
53: end
54:
55: %...Determine geometrical properties
56: [A,Qx,Qy,Ix,Iy,Ixy,xbar,ybar]=prop(x,y);
57: %...Shift to centroidal axis
58: [Ixbar,Iybar,Ixybar]= ...
59:    shftprop(A,xbar,ybar,Ix,Iy,Ixy);
60: xs=x-xbar; ys=y-ybar; xxs=xx-xbar; yys=yy-ybar;
61:
62: %...Find the first moment of area
63: [D1,Q,D2,D3,D4,D5,D6,D7]=prop(xxs,yys);
64:
65: %...Compute shear flow
66: Q=abs(Q); q=(V/Ixbar)*Q;
67:
68: fprintf('\n\nShear Flow at Interface');
69: fprintf(  '\n-----------------------');
70: fprintf( ...
71:    '\n\nTable of Nodes Forming Cross Section');
72: fprintf( ...
73:    '\n\n  Node #          x               y');
74: for i=1:length(x)
75:    fprintf('\n %3.0f %12.5g  %12.5g', ...
76:            i,x(i),y(i));
77: end
```

6.3. Distribution of Shear Stresses Due to Bending

```
78:  fprintf('\n\nGeometrical Properties');
79:  fprintf(   '\n   A:      %g',A);
80:  fprintf(   '\n   Qx:     %g',Qx);
81:  fprintf(   '\n   Qy:     %g',Qy);
82:  fprintf(   '\n   Ix:     %g',Ix);
83:  fprintf(   '\n   Iy:     %g',Iy);
84:  fprintf(   '\n   Ixy:    %g',Ixy);
85:  fprintf(   '\n   x-bar: %g',xbar);
86:  fprintf(   '\n   y-bar: %g',ybar);
87:  fprintf('\n\nGeometrical Properties About ');
88:  fprintf(     'Centroidal Axis');
89:  fprintf(   '\n   Ix-bar:  %g',Ixbar);
90:  fprintf(   '\n   Iy-bar:  %g',Iybar);
91:  fprintf(   '\n   Ixy-bar: %g',Ixybar);
92:  fprintf( ...
93:     '\n\nTable of Nodes Forming Subarea');
94:  fprintf( ...
95:     '\n\n   Node #          x              y');
96:  for i=1:length(xx)
97:     fprintf('\n  %3.0f %12.5g  %12.5g', ...
98:              i,xx(i),yy(i));
99:  end
100: fprintf('\n\nShear (V):        %g',V);
101: fprintf('\nFirst moment (Q): %g',Q);
102: fprintf('\nShear flow (q):   %g',q);
103: fprintf('\n\n');
104:
105: xp=[x(:);x(1)];   yp=[y(:);y(1)];
106: xxp=[xx(:);xx(1)]; yyp=[yy(:);yy(1)];
107: clf;
108:    fill(xxp,yyp,'y'); hold on
109:    plot(xp,yp,'-',xp,yp,'o');
110:    title('Cross Section Analyzed');
111:    xlabel('x'); ylabel('y');
112:    axis('equal'); hold off; drawnow;
113: % genprint('shrstr1');
```

Script File shrstr2

```
1: % Example: shrstr2
2: % ~~~~~~~~~~~~~~~~
3: % This example determines the shear stress
```

```matlab
 4: % distribution in a polygonal cross section.
 5: %
 6: % Data is defined in the declaration statements
 7: % below, where:
 8: %
 9: % x,y - vectors containing the coordinates for
10: %       the polygon nodes.  CCW fashion for
11: %       positive contributions, CW fashion
12: %       for negative contributions.
13: % V   - shear on the cross section
14: % No_planes - number of horizontal planes to
15: %             use for clipping
16: %
17: % NOTE: To make the plotting perform
18: %       as desired an undocumented MATLAB
19: %       capability has been used (Renderlimits).
20: %       MathWorks does not promise that it
21: %       will be supported in the future.
22: %       Renderlimits allows the two side-by-
23: %       side plots to maintain the same
24: %       y-axes.
25: %
26: % User m functions required:
27: %     prop, shftprop, genprint, clip, inout,
28: %     intrsect, genarea, findt
29: %-------------------------------------------------
30:
31: clear;
32: %...Input definitions
33: Problem=3;
34: if Problem == 1
35:    %...Rectangle, bottom left corner at (0,0)
36:    x=[ 0  1  1  0]; y=[ 0  0  5  5];
37:    V=1; No_planes=10;
38: elseif Problem == 2
39:    %...T-shaped cross section
40:    x=[0 .5 .5 2.25 2.25 -1.75 -1.75 0];
41:    y=[0 0 2 2 2.5 2.5 2 2];
42:    V=1.5; No_planes=20;
43: elseif Problem == 3
44:    %...I beam
45:    x=[3 3 0.5 0.5 3 3 -3 -3 -.5 -.5 -3 -3];
46:    y=[-5 -4.5 -4.5 4.5 4.5 5 5 ...
47:       4.5 4.5 -4.5 -4.5 -5];
```

6.3. Distribution of Shear Stresses Due to Bending

```
48:    V=2000; No_planes=80;
49: end
50:
51: %...Initialize
52: No_pts=length(x); n1=No_planes+1;
53: Q=zeros(1,n1); q=zeros(1,n1);
54: t=zeros(1,n1); tau=zeros(1,n1);
55:
56: %...Determine geometrical properties
57: [A,Qx,Qy,Ix,Iy,Ixy,xbar,ybar]=prop(x,y);
58: %...Shift to centroidal axis
59: [Ixbar,Iybar,Ixybar]= ...
60:    shftprop(A,xbar,ybar,Ix,Iy,Ixy);
61: xs=x-xbar; ys=y-ybar;
62: %...Extreme coordinates
63: ymax=max(ys); ymin=min(ys);
64:
65: %...Horizontal planes to find shear stresses at
66: y_shear=linspace(ymin,ymax,n1);
67:
68: %...Calculate shear at each plane
69: No_pts_clip=No_pts; x_clip=xs; y_clip=ys;
70: for i=2:No_planes
71:    %...Generate the clipped area
72:    [No_pts_clip,x_clip,y_clip]=genarea( ...
73:       No_pts_clip,x_clip,y_clip,y_shear(i));
74:    %...Find the first moment of area
75:    [D1,Q(i),D2,D3,D4,D5,D6,D7]= prop( ...
76:       x_clip,y_clip);
77:    %...Determine the t at the plane of interest
78:    [t(i)]=findt(x_clip,y_clip,y_shear(i));
79: end
80: %...Compute shear flow and stress
81: q=(V/Ixbar)*Q;
82: n=n1-1; tau(2:n)=q(2:n)./t(2:n);
83:
84: %...Shift coordinates to original system
85: y_shear=y_shear+ybar;
86:
87: fprintf('\n\nShear Stresses in Cross Section');
88: fprintf(  '\n--------------------------------');
89: fprintf( ...
90:    '\n\nTable of Nodes Forming Cross Section');
91: fprintf( ...
```

```
 92:           '\n\n   Node #            x                y');
 93:    for i=1:No_pts
 94:       fprintf('\n   %3.0f %12.5g  %12.5g', ...
 95:               i,x(i),y(i));
 96:    end
 97:    fprintf('\n\nGeometrical Properties');
 98:    fprintf(  '\n   A:      %g',A);
 99:    fprintf(  '\n   Qx:     %g',Qx);
100:    fprintf(  '\n   Qy:     %g',Qy);
101:    fprintf(  '\n   Ix:     %g',Ix);
102:    fprintf(  '\n   Iy:     %g',Iy);
103:    fprintf(  '\n   Ixy:    %g',Ixy);
104:    fprintf(  '\n   x-bar:  %g',xbar);
105:    fprintf(  '\n   y-bar:  %g',ybar);
106:    fprintf('\n\nGeometrical Properties About ');
107:    fprintf(     'Centroidal Axis');
108:    fprintf(  '\n   Ix-bar:  %g',Ixbar);
109:    fprintf(  '\n   Iy-bar:  %g',Iybar);
110:    fprintf(  '\n   Ixy-bar: %g',Ixybar);
111:    fprintf('\n\nShear (V): %g',V);
112:    for i=1:No_planes+1
113:       fprintf('\n\nClipping plane number %g',i);
114:       fprintf('\n   y:                  %g', ...
115:               y_shear(i));
116:       fprintf('\n   First moment (Q): %g',Q(i));
117:       fprintf('\n   Thickness:        %g',t(i));
118:       fprintf('\n   Shear flow (q):   %g',q(i));
119:       fprintf('\n   Shear stress:     %g',tau(i));
120:    end
121:    fprintf('\n\n');
122:
123:    xp=[x(:);x(1)]; yp=[y(:);y(1)];
124:    clf;
125:    ax=subplot(1,2,1);
126:       fill(xp,yp,'y'); hold on
127:       plot(xp,yp,'-',xp,yp,'o');
128:       title('Cross Section Analyzed');
129:       xlabel('x'); ylabel('y');
130:       axis('equal'); hold off; drawnow;
131:       Rlimits=get(ax,'RenderLimits');
132:    subplot(1,2,2);
133:       plot(tau,y_shear,'-')
134:       title( ...
135:          'Shear Stress Distribution');
```

6.3. Distribution of Shear Stresses Due to Bending

```
136:    xlabel('Shear Stress'); ylabel('y');
137:    set(gca,'YLim',[Rlimits(3),Rlimits(4)])
138:    drawnow;
139: % genprint('shrstr2');
```

Function genarea

```
 1: function [No_pts_clip,x_clip,y_clip]= ...
 2:          genarea(No_pts_orig,x_orig, ...
 3:          y_orig,y_clip_plane)
 4: %
 5: % [No_pts_clip,x_clip,y_clip]= ...
 6: %        genarea(No_pts_orig,x_orig, ...
 7: %        y_orig,y_clip_plane)
 8: %~~~~~~~~~~~~~~~~~~~~~~~~~~~~~~~~~~~~~~
 9: % This function generates a new polygon
10: % by clipping the reference polygon at
11: % the prescribed horizontal plane.
12: %
13: % No_pts_orig - number of coordinates defining
14: %               the reference polygon
15: % x_orig      - vector of x coordinates in
16: %               the reference polygon
17: % y_orig      - vector of y coordinates in
18: %               the reference polygon
19: % y_clip_plane- the y coordinate of the
20: %               horizontal clipping plane
21: %
22: % No_pts_clip - number of coordinates defining
23: %               the clipped polygon
24: % x_clip      - vector of x coordinates in
25: %               the clipped polygon
26: % y_clip      - vector of y coordinates in
27: %               the clipped polygon
28: %
29: % User m functions called:  clip
30: %---------------------------------------
31:
32: Epsilon=1e-8;
33:
34: %...Note: Define the clipping plane using the
35: %...      normal form:
```

```
36: %...
37: %...          alpha     beta*x    gamma*y
38: %...          -----  +  ------ +  ------- = 0
39: %...            d         d         d
40: %...
41: %...          where:
42: %...
43: %...          d=SIGN(beta)*SQRT(beta^2+gamma^2)
44: %...
45: Nform(1)=y_clip_plane; % perpendicular distance
46: Nform(2)=0;            % run of normal vector
47: Nform(3)=-1;           % rise of normal vector
48:
49: x_clip=x_orig; y_clip=y_orig;
50: NoClipPts=No_pts_orig;
51:
52: %...Clip the polygon
53: [NoNewPts,IsetClipPoly,NoInterpPts, ...
54:    x_clip,y_clip]=clip(x_clip,y_clip,NoClipPts, ...
55:    Nform,Epsilon);
56:
57: %...Clip complete, reorder arrays
58: Loop=NoClipPts+NoInterpPts;
59: xtmp=x_clip; ytmp=y_clip;
60: clear x_clip y_clip;
61: for j=1:NoNewPts
62:    itmp=IsetClipPoly(j);
63:    x_clip(j)=xtmp(itmp); y_clip(j)=ytmp(itmp);
64: end
65: No_pts_clip=NoNewPts;
```

Function findt

```
1: function [t]=findt(x,y,yplane)
2: %
3: % [t]=findt(x,y,yplane)
4: % ~~~~~~~~~~~~~~~~~~~~~
5: % This function calculates the thickness
6: % at a plane previously used to clip a
7: % polygon.
8: %
9: % x        - vector of area x coordinates
```

```
10: %   y       - vector of area y coordinates
11: %   yplane  - the y coordinate which defines
12: %             the elevation where the thickness
13: %             is found.  This should the plane
14: %             previously used by the clipping
15: %             algorithm for clipping.
16: %
17: %   t       - thickness
18: %
19: % User m functions called:  none
20: %-----------------------------------------------
21:
22: Epsilon=1.0e-8; % Value for determining zero
23: t=0; Npts=length(x);
24: for i=1:Npts
25:    j=i+1;
26:    if i == Npts, j=1; end;
27:    testi=abs(y(i)-yplane);
28:    testj=abs(y(j)-yplane);
29:    if (testi < Epsilon) & (testj < Epsilon)
30:       t=t+(x(j)-x(i));
31:    end
32: end
```

6.4 Bending of a Curved Beam with a Polygonal Cross Section

The analysis of stresses in curved beams is encountered when designing structural members such as grappling hooks. In this section traditional methods for calculating the bending stresses in curved beams are enhanced to incorporate the analysis of beams having symmetric polygonal cross sections. The mathematical relationships necessary to calculate the distribution of bending stresses in a curved beam are presented and a program to calculate the bending stresses is developed and discussed.

6.4.1 Bending Stress in a Curved Beam

The cross section shown in Figure 6.13 develops a linear variation of displacement when subjected to a bending load about the x-axis, M_x[8]. The point C shown in the figure represents the location of the center of curvature for the curved beam and ρ is the radius of curvature. The displacement normal to the section for any elevation

---

[8] Oden [33] has formulated the more complex solution of a cross section subjected to both M_x and M_y.

y can be expressed by the linear equation

$$\delta_y = a + by$$

where a and b are coefficients to be determined and the position y for any plane on the cross section is measured from the center of curvature axis[9]. The normal strain at any plane y can subsequently be written as

$$\epsilon_y = \frac{\delta_y}{y}$$

and when this equation is combined with the previous expression the strain becomes

$$\epsilon_y = \frac{1}{y}(a + by) = \frac{a}{y} + b$$

The normal stress is related to the strain by Young's modulus, or

$$\sigma_y = E\epsilon_y = \left(\frac{a}{y} + b\right) E = \frac{\alpha}{y} + \beta$$

where α and β are unknown constants for the bending stress equation. For the case of pure bending ($P = 0$) the sum of the internal forces on the cross section must be zero, or

$$\int\int \sigma \, dx \, dy = 0$$

and substituting the linear bending stress equation produces

$$\alpha \int\int \frac{dx \, dy}{y} + \beta \int\int dx \, dy = 0$$

or

$$\alpha K + \beta A = 0$$

where

$$K = \int\int \frac{dx \, dy}{y} \qquad A = \int\int dx \, dy$$

and K represents a new geometrical property of the cross section.

The moment produced by the internal forces on the cross section must balance the external moment on the cross section, or

$$M_x = \int\int y \, \sigma \, dx \, dy$$

---

[9] Many texts present the relationships for curved beams with y measured from the neutral axis. The authors have selected a coordinate system measured from the center of curvature to simplify the development of the equations for polygonal cross sections.

6.4. Bending of a Curved Beam with a Polygonal Cross Section

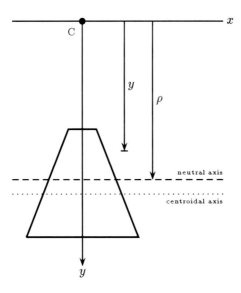

Figure 6.13. Cross Section of Curved Beam

When the linear stress equation is substituted into this relationship it becomes

$$M_x = \alpha \int\int dx\, dy + \beta \int\int y\, dx\, dy$$

or

$$\alpha A + \beta Q_x = M_x$$

where

$$Q_x = \int\int y\, dx\, dy$$

defines the first moment of the area. The two equations produced from statics can now be solved simultaneously to uniquely determine the two unknown constants which yields

$$\alpha = \frac{M_x A}{A^2 - Q_x K} \qquad \beta = -\frac{M_x K}{A^2 - Q_x K}$$

6.4.2 Algorithm for Geometry Integral

Chapter 4 presented the development of algorithms to determine the geometrical properties A, $\bar{x}$, $\bar{y}$, Q_x, Q_y, I_x, I_y, and I_{xy} for the case of an arbitrary polygon. The program developed in this section, which extends this set of geometrical properties, requires the evaluation of the integral[10]

$$K = \iint \frac{dx\,dy}{y}$$

This integral can be evaluated for an n-sided polygon, whose corner nodes have coordinates (x_i, y_i), using a side-by-side summation taking i from 1 to n and using $j = i + 1$ (except for $i = n$ when j is taken as 1), or

$$K = \sum_{i=1}^{n} S_i$$

where

$$S_i = \begin{cases} 0 & y_j = y_i, \quad horizontal \\ x_i \ln\left(\frac{y_j}{y_i}\right) & x_j - x_i = 0, \quad vertical \\ \left(\frac{A_i}{y_j - y_i}\right) \ln\left(\frac{y_j}{y_i}\right) + (x_j - x_i) & general\ side \end{cases}$$

and

$$A_i = (x_i y_j - x_j y_i)$$

6.4.3 Traditional Forms for Curved Beam Analysis

The equation for the bending stress in a curved beam is written in many texts as

$$\sigma = \frac{M(y - \rho)}{AEy} = \frac{M}{AE} - \frac{1}{y}\left(\frac{M\rho}{Ae}\right)$$

where y is measured from the neutral axis of the cross section, e is the distance between the neutral axis and the centroidal axis (or $e = \bar{y} - \rho$), and ρ is the radius of curvature for the cross section. The radius of curvature can be obtained by evaluating the relationship

$$\rho = \frac{\iint dx\,dy}{\iint \frac{dx\,dy}{\rho - y}}$$

The numerator of this expression is simply the area of the cross section and the denominator reflects a new geometrical property for the cross section (which was presented in the previous subsection). This expression for bending stress is equivalent to the previously developed equation

$$\sigma_y = \frac{\alpha}{y} + \beta$$

---

[10] The mathematical development can be performed using line integrals and is similar to the methods used to find geometrical properties of polygons which was presented by Wilson and Turcotte [54].

6.4. Bending of a Curved Beam with a Polygonal Cross Section

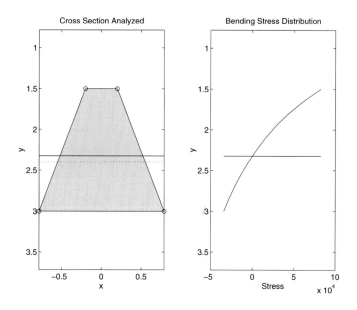

Figure 6.14. Bending Stress Distribution

6.4.4 Program to Calculate Normal Stress in a Curved Beam

A program was written to calculate and plot the distribution of bending stresses in a curved beam having a polygonal cross section. The output for a trapezoidal cross section is included in Section 6.4.6. A description of the code is summarized in Table 6.4. Figure 6.14 provides the graphical results from the analysis. The cross section is plotted in the left graph with the centroidal axis designated by a dotted line and the neutral axis indicated by a solid line. The graph on the right shows the distribution of the normal stresses due to pure bending. It should be noted that the bending stress distribution is not linear, but is hyperbolic.

6.4.5 Exercises

1. Modify the program for curved beam analysis to calculate the bending stresses at the extreme (top and bottom) fibers of a cross section for a range of values for the radius of curvature. This can be achieve by offsetting the y coordinates which define the cross section by a prescribed set of values. Graph the relationship for stress versus radius of curvature. How does this compare with the equation for bending stress in a straight beam (i.e. $\sigma = My/I$)?

Routine	Line	Operation
cbeamex		script file to execute program.
	34	select a problem.
	35-51	define the input parameters.
	54-55	initialize some parameters.
	58-59	define the planes to determine the stress on.
	62-66	calculate the geometrical properties and two length measures.
	70	calculate stress equation constants.
	71	calculate the bending stress at each plane.
	72-104	output the results.
	106-129	plot results.
cbprop		function which calculates the geometrical properties.
	22	expand coordinate arrays to close polygon.
	23-36	loop on each side of polygon.
	27-35	calculate the contribution for the new geometrical property.

Table 6.4. Description of Code in Example **cbeamex**

6.4.6 Program Output and Code

Output from Example cbeamex

```
Stresses in Cross Section of Curved Beam
----------------------------------------

Table of Nodes Forming Cross Section

    Node #         x              y
       1        -0.785            3
       2         0.785            3
       3         0.2              1.5
       4        -0.2              1.5

Geometrical Properties
   A:      1.4775
   Qx:     3.54375
```

6.4. Bending of a Curved Beam with a Polygonal Cross Section

```
    K:     0.636277
    y-bar: 2.39848
    e:     0.0763741
    rho:   2.3221

Bending moment M:          -17000

Stress Equation Coefficients
   alpha: 349830
   beta:  -150652

Plane number 1
   y:                       1.5
   Bending stress:          82567.7
Plane number 2
   y:                       1.65
   Bending stress:          61365.9
Plane number 3
   y:                       1.8
   Bending stress:          43697.7
Plane number 4
   y:                       1.95
   Bending stress:          28747.8
Plane number 5
   y:                       2.1
   Bending stress:          15933.5
Plane number 6
   y:                       2.25
   Bending stress:          4827.77
Plane number 7
   y:                       2.4
   Bending stress:          -4889.73
Plane number 8
   y:                       2.55
   Bending stress:          -13464
Plane number 9
   y:                       2.7
   Bending stress:          -21085.5
Plane number 10
   y:                       2.85
   Bending stress:          -27904.8
Plane number 11
   y:                       3
   Bending stress:          -34042.2
```

Script File cbeamex

```
 1: % Example:cbeamex
 2: % ~~~~~~~~~~~~~~~
 3: % This program calculates the stress
 4: % distribution in a curved beam having a
 5: % polygonal cross section.
 6: %
 7: % Data is defined in the declaration statements
 8: % below, where:
 9: %
10: % x,y - vectors containing the coordinates for
11: %       the polygon nodes.  CCW fashion for
12: %       positive contributions, CW fashion
13: %       for negative contributions.
14: % M   - moment on the cross section
15: % No_planes - number of planes to calculate
16: %       stresses at
17: %
18: % NOTE: To make the plotting perform
19: %       as desired an undocumented MATLAB
20: %       capability has been used (Renderlimits).
21: %       MathWorks does not promise that it
22: %       will be supported in the future.
23: %       Renderlimits allows the two side-by-
24: %       side plots to maintain the same
25: %       y-axes.
26: %
27: % User m functions required:
28: %       cbprop, genprint
29: %-------------------------------------------
30:
31: clear;
32: %...Input definitions
33: Problem=2;
34: if Problem == 1
35:    %...Rectangular
36:    x=[0    0.04  0.04 0];
37:    y=[0.04 0.04  0.1  0.1];
38:    M=-1162; No_planes=10;
39: elseif Problem == 2
40:    %...Trapezoidal
41:    x=[-1.57/2 1.57/2 0.4/2 -0.4/2];
```

6.4. Bending of a Curved Beam with a Polygonal Cross Section

```
42:    y=[3.0      3.0      1.5      1.5];
43:    M=-17000; No_planes=10;
44: elseif Problem == 3
45:    %...T-shape cross section
46:    x=[-40  40  40  10  10 -10 -10 -40];
47:    y=[ 30  30  50  50  90  90  50  50];
48:    x=x/1000; y=y/1000; % metric conversion
49:    M=940.5; No_planes=10;
50: end
51:
52: %...Initialize
53: No_pts=length(x); n1=No_planes+1;
54: sigma=zeros(1,n1);
55:
56: %...Planes to find stresses at
57: ymin=min(y); ymax=max(y);
58: yi=linspace(ymin,ymax,n1);
59:
60: %...Geometrical properties of cross section
61: [A,Qx,K,ybar]=cbprop(x,y);
62:
63: %...Radius of curvature and the distance
64: %...between neutral axis and centroidal axis
65: rho=A/K; e=ybar-rho;
66:
67: %...Define stress equation constants and
68: %...solve for bending stresses
69: alpha=M*A/(A^2-Qx*K); beta=-M*K/(A^2-Qx*K);
70: sigma=alpha./yi+beta;
71:
72: fprintf( ...
73:    '\n\nStresses in Cross Section of Curved Beam');
74: fprintf( ...
75:    '\n----------------------------------------');
76: fprintf( ...
77:    '\n\nTable of Nodes Forming Cross Section');
78: fprintf( ...
79:    '\n\n  Node #          x               y');
80: for i=1:No_pts
81:    fprintf('\n %3.0f %12.5g  %12.5g', ...
82:            i,x(i),y(i));
83: end
84: fprintf('\n\nGeometrical Properties');
85: fprintf( '\n  A:       %g',A);
```

```
86:  fprintf(   '\n   Qx:      %g',Qx);
87:  fprintf(   '\n   K:       %g',K);
88:  fprintf(   '\n   y-bar:   %g',ybar);
89:  fprintf(   '\n   e:       %g',e);
90:  fprintf(   '\n   rho:     %g',rho);
91:  fprintf('\n');
92:  fprintf('\nBending moment M:          %g',M);
93:  fprintf('\n\nStress Equation Coefficients');
94:  fprintf(   '\n   alpha: %g',alpha);
95:  fprintf(   '\n   beta:  %g',beta);
96:  fprintf('\n');
97:
98:  for i=1:No_planes+1
99:    fprintf('\nPlane number %g',i);
100:   fprintf('\n  y:                      %g',yi(i));
101:   fprintf('\n  Bending stress:         %g', ...
102:     sigma(i));
103: end
104: fprintf('\n\n');
105:
106: xp=x; yp=y;
107: xp(No_pts+1)=xp(1); yp(No_pts+1)=yp(1);
108: xr1=[min(xp) max(xp)];
109: xr2=[min(sigma) max(sigma)];
110: yybar=[ybar ybar]; yrho=[rho rho];
111: clf;
112: ax=subplot(1,2,1);
113:   fill(xp,yp,'y'); hold on;
114:   plot(xp,yp,'-',xp,yp,'o');
115:   plot(xr1,yybar,'r:');
116:   plot(xr1,yrho,'g-');
117:   title('Cross Section Analyzed');
118:   xlabel('x'); ylabel('y');
119:   set(gca,'YDir','reverse');
120:   axis('equal'); hold off; drawnow;
121:   Rlimits=get(ax,'RenderLimits');
122: subplot(1,2,2);
123:   plot(sigma,yi,'-y',xr2,yrho,'g-');
124:   title('Bending Stress Distribution');
125:   xlabel('Stress'); ylabel('y');
126:   set(gca,'YDir','reverse');
127:   set(gca,'YLim',[Rlimits(3),Rlimits(4)]);
128:   drawnow;
129:   %genprint('cbtotal');
```

6.4. Bending of a Curved Beam with a Polygonal Cross Section

Function cbprop

```
function [A,Qx,K,ybar]=cbprop(x,y)
%
% [A,Qx,K,ybar]=cbprop(x,y)
% ~~~~~~~~~~~~~~~~~~~~~~~~~
% This function computes the geometrical
% properties for a polygon to use with
% curved beam analysis.
%
% x       - vector of x coordinates
% y       - vector of x coordinates
%
% A       - area of polygon
% Qx      - first moment of area about
%           x-axis
% K       - integral of (dx*dy)/y
% ybar    - y distance to centroid
%
% User m functions called: none
%-------------------------------------------

no_pts=length(x); A=0; K=0; Qx=0;
x=[x x(1)]; y=[y y(1)];
for i=1:no_pts
  xi=x(i); yi=y(i); xj=x(i+1); yj=y(i+1);
  Ai=xi*yj-xj*yi; A=A+Ai;
  Qx=Qx+(yi+yj)*Ai;
  %...(dx*dy)/y integral
  if (yj-yi) == 0
    %...horizontal line, Qi=0;
  elseif (xj-xi) == 0
    K=K+xi*log(yj/yi); %...vertical line
  else
    %...general line
    K=K+Ai/(yj-yi)*log(yj/yi)+(xj-xi);
  end
end
A=0.5*A; Qx=Qx/6; ybar=Qx/A;
%...Reverse signs if necessary (when nodes
%     traversed clockwise)
if A < 0, A=-A; Qx=-Qx; K=-K; end
```

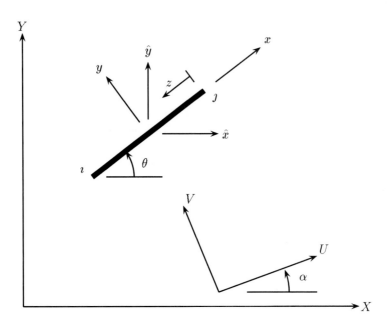

Figure 6.15. General Member of Thin-Walled Cross Section

6.5 Shear Center for Open Thin-Walled Members

The *shear center* for open thin-walled members represents the location at which a single transverse shear force may be placed such that no twisting moment is developed in the cross section. The shear center generally does not coincide with the centroid of the cross section and does not have to be a point on the cross section. In this section a program is presented[11] which determines the shear center for open thin-walled members.

6.5.1 Geometrical Properties

The solution of a multi-segment cross section can be achieved by developing a general method to analyze individual segments and then aggregating their contributions. Figure 6.15 shows a general segment from a cross section. A local coordinate system, (x, y), and a global system, (X, Y), are employed in the development. The

---

[11] Paz et al. [35] originally published a FORTRAN program which analyzed open and closed section thin-walled cross sections.

6.5. Shear Center for Open Thin-Walled Members

centroid of an individual segment and the angle it forms with the global system is

$$\bar{X}_k = \frac{(X_i)_k + (X_j)_k}{2} \qquad \bar{Y}_k = \frac{(Y_i)_k + (Y_j)_k}{2}$$

$$\theta_k = \tan^{-1}\left(\frac{Y_i - Y_j}{X_i - X_j}\right)_k$$

where i and j represent the opposite ends of the segment and k represents the k-th segment. The centroid of the entire cross section in the global coordinate system can be determined by summing the contributions from each segment, or

$$\bar{X} = \frac{\sum \bar{X}_k A_k}{\sum A_k} \qquad \bar{Y} = \frac{\sum \bar{Y}_k A_k}{\sum A_k} \qquad A_k = \ell_k t_k$$

where the area of a segment is A_k, the length of a segment is ℓ_k, and the thickness of a segment is t_k. The inertia properties for a segment in the local (x, y) coordinate system can be written as

$$(I_x)_k = \frac{t_k^3 \ell_k}{12} \qquad (I_y)_k = \frac{t_k \ell_k^3}{12} \qquad (I_{xy})_k = 0$$

These properties can be transformed by rotation to axes parallel to the global coordinate system which yields

$$(\hat{I}_x)_k = (I_x)_k \cos^2 \theta_k + (I_y)_k \sin^2 \theta_k$$

$$(\hat{I}_y)_k = (I_x)_k + (I_y)_k - (\hat{I}_x)_k$$

$$(\hat{I}_{xy})_k = -\frac{1}{2}[(I_x)_k - (I_y)_k] \sin 2\theta_k$$

The parallel-axis theorem can be employed to translate the contributions to the centroidal axes, or

$$I_X = \sum [(\hat{I}_x)_k + A_k(\bar{Y}_k - \bar{Y})^2]$$

$$I_Y = \sum [(\hat{I}_y)_k + A_k(\bar{X}_k - \bar{X})^2]$$

$$I_{XY} = \sum [A_k(\bar{X}_k - \bar{X})(\bar{Y}_k - \bar{Y})]$$

The angle to the principal inertia axes is

$$2\alpha = \tan^{-1}\left(\frac{2I_{XY}}{I_Y - I_X}\right)$$

and the corresponding values of inertia are

$$I_U = \frac{I_X + I_Y}{2} + \left(\frac{I_X - I_Y}{2}\right)\cos(2\alpha) - I_{XY}\sin(2\alpha)$$

$$I_V = I_X + I_Y - I_U$$

Finally, the coordinates of the right end of each member will be represented in the principal inertia system as

$$(u_j)_k = (X_j)_k \cos\alpha + (Y_j)_k \sin\alpha$$
$$(v_j)_k = -(X_j)_k \sin\alpha + (Y_j)_k \cos\alpha$$

where the location of the right end in the global system is

$$(X_j)_k = \bar{X}_k - \bar{X} + \frac{\ell_k}{2}\cos\theta_k \qquad (Y_j)_k = \bar{Y}_k - \bar{Y} + \frac{\ell_k}{2}\sin\theta_k$$

6.5.2 Shear Center

The shear stress at any point in a cross section (relative to the principal inertia axes) can written as

$$\tau = \frac{VQ}{It}$$

where V is the component of shear force parallel to the principal axis, I is the principal moment of inertia for the entire cross section for an axis perpendicular to V, t is the thickness of the cross section for the point considered, and Q is the first moment with respect to the principal axis. The value for Q at any point in a thin-walled open cross section can be determined by

$$Q = \sum Q_{\text{complete segments}} + Q_{\text{partial segment}}$$

The value for the first moment anywhere within a segment about each axis can be expressed as

$$[(Q_z)_k]_U = A_z \bar{y}_z = z t_k \left[(v_j)_k - \frac{z\sin(\theta_k - \alpha)}{2}\right]$$

$$[(Q_z)_k]_V = A_z \bar{x}_z = z t_k \left[(u_j)_k - \frac{z\sin(\pi/2 - \theta_k + \alpha)}{2}\right]$$

and for a complete section as

$$(Q_u)_k = \ell_k t_k (v_j)_k - \left(\frac{t_k \ell_k^2}{2}\right)\sin(\theta_k - \alpha)$$

$$(Q_v)_k = \ell_k t_k (u_j)_k - \left(\frac{t_k \ell_k^2}{2}\right)\sin(\pi/2 - \theta_k + \alpha)$$

The shear force which develops in the partial segment can be written by considering the contributions of shear along the length of the segment, or

$$[(F_z)_k]_x = \int^z t_k \tau \, dz$$

6.5. Shear Center for Open Thin-Walled Members

$$= \int^z t_k \left(\frac{VQ}{It}\right)_k dz$$

$$= \frac{V}{I_U} \int^z \left[zt_k(v_j)_k - \frac{z^2 t_k}{2}\sin(\theta_k - \alpha) + \sum (Q_u)_{\text{complete segments}}\right]$$

and the contribution perpendicular to the length of the segment, or

$$[(F_z)_k]_y = \frac{V}{I_V} \int^z \left[zt_k(u_j)_k - \frac{z^2 t_k}{2}\sin\left(\frac{\pi}{2} - \theta_k + \alpha\right) + \sum (Q_v)_{\text{complete segments}}\right]$$

The shear force developed in complete sections can be written as

$$(F_k)_x = \frac{V}{I_U}\left[\frac{\ell_k^2 t_k (v_j)_k}{2} - \frac{\ell_k^3 t_k \sin(\theta - \alpha)}{6} + \ell_k \sum (Q_u)_{\text{complete segments}}\right]$$

and

$$(F_k)_y = \frac{V}{I_V}\left[\frac{\ell_k^2 t_k (u_j)_k}{2} - \frac{\ell_k^3 t_k \sin\left(\frac{\pi}{2} - \theta_k + \alpha\right)}{6} + \ell_k \sum (Q_v)_{\text{complete segments}}\right]$$

The location of the shear center in the principal axes can be determined by

$$e_u = \sum (F_k)_x \sin(\theta_k - \alpha)(u_j)_k - \sum (F_k)_x \cos(\theta_k - \alpha)(v_j)_k$$

$$e_v = \sum (F_k)_y \sin\left(\frac{\pi}{2} - \theta_k + \alpha\right)(v_j)_k -$$

$$\sum (F_k)_y \cos\left(\frac{\pi}{2} - \theta_k + \alpha\right)(u_j)_k$$

and realizing that the magnitude of V does not influence the location of the shear center. Finally, the location of the shear center can be expressed in the original global coordinate system as

$$e_X = \bar{X} + e_u \cos\alpha - e_v \sin\alpha$$
$$e_Y = \bar{Y} + e_u \sin\alpha + e_v \cos\alpha$$

6.5.3 Program to Calculate the Shear Center

The program to calculate the shear center of an open thin-walled cross section is summarized in Table 6.5. The output from an example problem is provided in Section 6.5.5. The program demonstrates the capability to implement a sequence of complex relationships in a concise fashion. Approximately one hundred statements are required (excluding input and output statements) to determine the shear center of an open thin-walled cross section.

6.5.4 Exercises

1. Write a program which determines the shear stress at any specified point in the cross section.

2. Using the paper by Paz et al. [35] as a guide, enhance the program included in this section to analyze both open and closed thin-walled cross sections.

Routine	Line	Operation
shcenter		script file to execute program.
	35	select one of the example problems.
	36-69	define the input parameters.
	70-71	initialize some parameters.
	73-81	determine the centroid for each member in the global coordinate system.
	85	determine the area of each member.
	86-87	determine the first moment of each member in the global coordinate system.
	89	calculate the total area of the cross section.
	90	calculate the first moments of the cross section.
	91	calculate the centroid of the cross section.
	96-97	determine the moment of inertia for each member in the local coordinate sytem.
	100-103	rotate the inertia values to coordinate system parallel to global coordinate system.
		continued on next page

6.5. Shear Center for Open Thin-Walled Members

Routine	Line	Operation
		continued from previous page
	105-110	determine the moment inertia for the cross section in the global coordinate system using the parallel-axis theorem.
	113	calculate the angle to the principal inertia axes.
	114-116	calculate the principal inertia values.
	120-127	find the coordinate of the right end of each member in the principal coordinate system.
	129-135	determine the first moment for each member about the principal axes.
	138-147	calculate the accumulated values of first moment for each member which are be contributed by preceding members attached to the member.
	151-167	determine the location of the shear center in the principal axis system.
	170-171	calculate the location of the shear center in the original global coordinate system.
	173-253	output the results.
	255-261	plot the cross section.

Table 6.5. Description of Code in Example **shcenter**

6.5.5 Program Output and Code

Output from Example shcenter

```
Shear Center of Open Thin Walled Sections
-----------------------------------------

Table of Nodes Forming Cross Section

  Node #          x               y
    1            0.5             0.97
    2            0.03            0.97
    3            0.03            0.03
    4            1               0.03

Table of Member Properties

  Member #        L               t
    1            0.47            0.06
```

2	0.94	0.06
3	0.97	0.06

Member Connectivity

Member #	Start Node	End Node
1	1	2
2	2	3
3	3	4

Connectivity Sequence

Member #	Preceding Members	
1		
2	1	
3	1	2

Member Geometrical Properties

  Member # 1
    Local (x,y)
      A: 0.0282
      Ix: 8.46e-006
      Iy: 0.000519115
      x-bar: 0.265
      y-bar: 0.97
    Rotated local (x,y)
      Angle: 0 degs
      Ix-hat: 8.46e-006
      Iy-hat: 0.000519115
      Ixy-hat: 0
    Global (x,y)
      Qx: 0.027354
      Qy: 0.007473
      Ix: 0.00913016
      Iy: 0.000521438
      Ixy: -0.000145559

  Member # 2
    Local (x,y)
      A: 0.0564
      Ix: 0.00001692
      Iy: 0.00415292
      x-bar: 0.03

6.5. Shear Center for Open Thin-Walled Members

```
        y-bar:    0.5
      Rotated local (x,y)
        Angle:    90 degs
        Ix-hat:   0.00415292
        Iy-hat:   0.00001692
        Ixy-hat:  0.002068
      Global (x,y)
        Qx:       0.0282
        Qy:       0.001692
        Ix:       0.00470279
        Iy:       0.00337683
        Ixy:     -0.00135923

  Member # 3
    Local (x,y)
      A:        0.0582
      Ix:       0.00001746
      Iy:       0.00456336
      x-bar:    0.515
      y-bar:    0.03
    Rotated local (x,y)
      Angle:    180 degs
      Ix-hat:   0.00001746
      Iy-hat:   0.00456336
      Ixy-hat:  3.40866e-035
    Global (x,y)
      Qx:       0.001746
      Qy:       0.029973
      Ix:       0.00803942
      Iy:       0.00794156
      Ixy:     -0.00520574

Total (Global)
  A:        0.1428
  x-bar:    0.274076
  y-bar:    0.401261
  Qx:       0.0573
  Qy:       0.039138
  Ix:       0.0218724
  Iy:       0.0118398
  Ixy:     -0.00671053

Geometrical Properties About Principal Axis
  Angle:   -63.3895 degs
```

Iu: 0.0084779
Iv: 0.0252343

Location of Shear Center
  Original axes
    x: -0.179456
    y: 0.185684
  Principal axes
    u: -0.0104065
    v: -0.502051

Script File shcenter

```
 1: % Example: shcenter
 2: % ~~~~~~~~~~~~~~~~
 3: % This example determines the shear center of
 4: % open thin walled sections.
 5: %
 6: % Data is defined in the declaration statements
 7: % below, where:
 8: %
 9: %------------------------------------------------
10: %  x,y        - vectors containing the coordinates
11: %                for the nodes of the cross section.
12: %  Length     - vector containing length of each
13: %                member of cross section.
14: %  Thick      - vector containing the thickness
15: %                of each member of cross section
16: %  Connect    - connectivity matrix.  Each row
17: %                contains the starting node and
18: %                ending node which define the
19: %                member.
20: %  NoPrecede  - a vector which indicates how
21: %                many other members are attached
22: %                to this member to reach a free
23: %                end of the cross section.
24: %  Precede    - a matrix containing the list
25: %                of members attached to this
26: %                member. Each row has NoPrecede
27: %                entries and is the corresponding
28: %                member number which is attached.
29: %
30: % User m functions required:
```

6.5. Shear Center for Open Thin-Walled Members

```
31: %      genprint
32: %-----------------------------------------------
33:
34: clear;
35: Problem=1;
36: if Problem == 1
37:    %...From Paz
38:    x=[0.5  0.03 0.03 1    ];
39:    y=[0.97 0.97 0.03 0.03];
40:    %...for each member
41:    Connect=[1 2; 2 3; 3 4];
42:    Length=[0.47 0.94 0.97];
43:    Thick= [0.06 0.06 0.06];
44:    NoPrecede=[0 1 2];
45:    Precede=[0 0 0; 1 0 0; 1 2 0];
46: elseif Problem == 2
47:    %..Schaum 12.4 (Nash)
48:    x=[ 3  1    0    0   1 3];
49:    y=[-3 -3 -1.5  1.5   3 3];
50:    %...for each member
51:    Connect=[1 2; 2 3; 3 4; 4 5; 5 6];
52:    d=sqrt(1+1.5^2);
53:    Length=[2 d 3 d 2];
54:    Thick= [.1 .1 .1 .1 .1];
55:    NoPrecede=[0 1 2 3 4];
56:    Precede=[0 0 0 0 0; 1 0 0 0 0; 1 2 0 0 0; ...
57:             1 2 3 0 0; 1 2 3 4 0];
58: elseif Problem == 3
59:    %..Schaum 12.3 (Nash)
60:    x=[50  50    0   0 50 50];
61:    y=[ 0 -50  -50  50 50  0];
62:    %...for each member
63:    Connect=[1 2; 2 3; 3 4; 4 5; 5 6];
64:    Length=[50 50 100 50 50];
65:    Thick= [1 1 1 1 1];
66:    NoPrecede=[0 1 2 3 4];
67:    Precede=[0 0 0 0 0; 1 0 0 0 0; 1 2 0 0 0; ...
68:             1 2 3 0 0; 1 2 3 4 0];
69: end
70: No_pts=length(x); NoMembers=length(Length);
71: raddeg=180/pi;
72:
73: %...Get centroid of each member and
74: %...the angle of each member (global system)
```

```
75: for k=1:NoMembers
76:   i=Connect(k,1); j=Connect(k,2);
77:   xi=x(i); yi=y(i); xj=x(j); yj=y(j);
78:   XbarMember(k)=(xi+xj)/2;
79:   YbarMember(k)=(yi+yj)/2;
80:   Theta(k)=atan2(yi-yj,xi-xj);
81: end
82:
83: %...Area, first moment, and centroid of member
84: %...(global system)
85: AreaMember=Length.*Thick;
86: QxMember=YbarMember.*AreaMember;
87: QyMember=XbarMember.*AreaMember;
88: %...For entire geometry (global system)
89: Area=sum(AreaMember);
90: Qx=sum(QxMember); Qy=sum(QyMember);
91: Xbar=Qy/Area; Ybar=Qx/Area;
92:
93: %...Inertia properties
94: %...(Ix,Iy) for each member for local
95: %...(x,y) axes (x along length axis)
96: IxMember=Thick.^3.*Length/12;
97: IyMember=Thick.*Length.^3/12;
98: %...Rotated to local (x,y) axes which is
99: %...parallel to global system
100: sinAng=sin(Theta).^2; cosAng=cos(Theta).^2;
101: IxHatMember=IxMember.*cosAng+IyMember.*sinAng;
102: IyHatMember=IxMember+IyMember-IxHatMember;
103: IxyHatMember=-0.5*(IxMember-IyMember).*sinAng;
104: %...Contribution to entire geometry using
105: %...parallel axis theorem
106: Ix=IxHatMember+AreaMember.*(YbarMember-Ybar).^2;
107: Iy=IyHatMember+AreaMember.*(XbarMember-Xbar).^2;
108: Ixy=AreaMember.*(YbarMember-Ybar).* ...
109:     (XbarMember-Xbar);
110: IX=sum(Ix); IY=sum(Iy); IXY=sum(Ixy);
111:
112: %...Principal axes
113: tmp=atan2(2*IXY,IY-IX); Alpha=0.5*tmp;
114: Iu=(IX+IY)/2+((IX-IY)/2)*cos(tmp)- ...
115:     IXY*sin(tmp);
116: Iv=IX+IY-Iu;
117:
118: %...Transform coordinates to principal axes
```

6.5. Shear Center for Open Thin-Walled Members

```
119:  %...system
120:  sinAng=sin(Theta)/2; cosAng=cos(Theta)/2;
121:  %...Right end of member
122:  Xr=XbarMember-Xbar+Length.*cosAng;
123:  Yr=YbarMember-Ybar+Length.*sinAng;
124:  %...Coordinates in principal axes system
125:  sinAng=sin(Alpha); cosAng=cos(Alpha);
126:  U=Xr*cosAng+Yr*sinAng;
127:  V=-Xr*sinAng+Yr*cosAng;
128:
129:  %...First moments about principal axes
130:  betaX=Theta-Alpha; betaX=sin(betaX);
131:  betaY=pi/2-(Theta-Alpha); betaY=sin(betaY);
132:  Qu=V.*Thick.*Length- ...
133:      Length.^2.*betaX.*Thick/2;
134:  Qv=U.*Thick.*Length- ...
135:      Length.^2.*betaY.*Thick/2;
136:
137:  %...Get contributions from previous members
138:  for k=1:NoMembers
139:     qsumU=0; qsumV=0;
140:     if NoPrecede(k) ~= 0
141:       for i=1:NoPrecede(k)
142:         m=Precede(k,i);
143:         qsumU=qsumU+Qu(m); qsumV=qsumV+Qv(m);
144:       end
145:     end
146:     QuPrev(k)=qsumU; QvPrev(k)=qsumV;
147:  end
148:
149:  %...Find measure for x-eccentricity in
150:  %...principal inertia axis
151:  beta=Theta-Alpha;
152:  sinbeta=sin(beta); cosbeta=cos(beta);
153:  F=(QuPrev.*Length+V.*Thick.* ...
154:     Length.^2/2-Thick.* ...
155:     Length.^3.*sinbeta/6)/Iu;
156:  mFx=F.*cosbeta.*V; mFy=F.*sinbeta.*U;
157:  eU=sum(mFy)-sum(mFx);
158:
159:  %...Find measure for y-eccentricity in
160:  %...principal inertia axis
161:  beta=pi/2-(Theta-Alpha);
162:  sinbeta=sin(beta); cosbeta=cos(beta);
```

```
163:  F=(QvPrev.*Length+U.*Thick.* ...
164:     Length.^2/2-Thick.* ...
165:     Length.^3.*sinbeta/6)/Iv;
166:  mFx=F.*cosbeta.*U; mFy=F.*sinbeta.*V;
167:  eV=sum(mFy)-sum(mFx);
168:
169:  %...Eccentricity in original global system
170:  xg=Xbar+eU*cos(Alpha)-eV*sin(Alpha);
171:  yg=Ybar+eU*sin(Alpha)+eV*cos(Alpha);
172:
173:  %...Output results
174:  fprintf('\n\nShear Center of Open Thin ');
175:  fprintf('Walled Sections');
176:  fprintf('\n----------------------------');
177:  fprintf('-------------');
178:  fprintf( ...
179:     '\n\nTable of Nodes Forming Cross Section');
180:  fprintf( ...
181:     '\n\n  Node #         x              y');
182:  for i=1:No_pts
183:     fprintf('\n %3.0f %12.5g  %12.5g', ...
184:             i,x(i),y(i));
185:  end
186:  fprintf('\n\nTable of Member Properties');
187:  fprintf( ...
188:     '\n\n  Member #       L              t');
189:  for i=1:NoMembers
190:     fprintf('\n %3.0f %12.5g  %12.5g', ...
191:             i,Length(i),Thick(i));
192:  end
193:  fprintf('\n\nMember Connectivity');
194:  fprintf( ...
195:     '\n\n  Member #  Start Node  End Node');
196:  for i=1:NoMembers
197:     fprintf('\n %3.0f   %10.0f %10.0f', ...
198:             i,Connect(i,1),Connect(i,2));
199:  end
200:  fprintf('\n\nConnectivity Sequence');
201:  fprintf( ...
202:     '\n\n  Member #  Preceding Members');
203:  for i=1:NoMembers
204:     fprintf('\n  %3.0f',i);
205:     for j=1:NoPrecede(i)
206:        fprintf('    %3.0f',Precede(i,j));
```

6.5. Shear Center for Open Thin-Walled Members

```
207:    end
208:  end
209:  fprintf('\n\nMember Geometrical Properties');
210:  for k=1:NoMembers
211:     fprintf('\n\n   Member # %g',k);
212:     fprintf('\n      Local (x,y)');
213:     fprintf('\n          A:       %g',AreaMember(k));
214:     fprintf('\n          Ix:      %g',IxMember(k));
215:     fprintf('\n          Iy:      %g',IyMember(k));
216:     fprintf('\n          x-bar:   %g',XbarMember(k));
217:     fprintf('\n          y-bar:   %g',YbarMember(k));
218:     fprintf('\n      Rotated local (x,y)');
219:     fprintf('\n          Angle:   %g',Theta(k)*raddeg);
220:     fprintf(' degs');
221:     fprintf('\n          Ix-hat:  %g',IxHatMember(k));
222:     fprintf('\n          Iy-hat:  %g',IyHatMember(k));
223:     fprintf('\n          Ixy-hat: %g',IxyHatMember(k));
224:     fprintf('\n      Global (x,y)');
225:     fprintf('\n          Qx:      %g',QxMember(k));
226:     fprintf('\n          Qy:      %g',QyMember(k));
227:     fprintf('\n          Ix:      %g',Ix(k));
228:     fprintf('\n          Iy:      %g',Iy(k));
229:     fprintf('\n          Ixy:     %g',Ixy(k));
230:  end
231:  fprintf('\n\nTotal (Global)')
232:  fprintf('\n   A:      %g',Area);
233:  fprintf('\n   x-bar:  %g',Xbar);
234:  fprintf('\n   y-bar:  %g',Ybar);
235:  fprintf('\n   Qx:     %g',Qx);
236:  fprintf('\n   Qy:     %g',Qy);
237:  fprintf('\n   Ix:     %g',IX);
238:  fprintf('\n   Iy:     %g',IY);
239:  fprintf('\n   Ixy:    %g',IXY);
240:  fprintf('\n\nGeometrical Properties About ');
241:  fprintf(   'Principal Axis');
242:  fprintf('\n   Angle: %g',Alpha*raddeg);
243:  fprintf(' degs');
244:  fprintf('\n   Iu:    %g',Iu);
245:  fprintf('\n   Iv:    %g',Iv);
246:  fprintf('\n\nLocation of Shear Center');
247:  fprintf(  '\n   Original axes');
248:  fprintf(  '\n      x: %g',xg);
249:  fprintf(  '\n      y: %g',yg);
250:  fprintf(  '\n   Principal axes');
```

```
251: fprintf( '\n      u: %g',eU);
252: fprintf( '\n      v: %g',eV);
253: fprintf('\n\n');
254:
255: clf;
256: plot(x,y,'-',x,y,'o',xg,yg,'+g');
257:    title('Original Geometry');
258:    xlabel('x'); ylabel('y');
259:    axis('equal'); tmp=legend('+g','  Shear Center ');
260:    axes(tmp); drawnow;
261:    %genprint('shcent');
```

Chapter 7

Flexural Analysis and Deflection of Beams Using Discontinuity Functions

In this chapter discontinuity functions are employed to calculate the shear, moment, rotation[1], and deflection at points along simply supported beams of constant cross section. A review of discontinuity functions is followed by a short discussion of their application to beam analysis. The development of two computer programs is also presented. The first program employs an approximate method to calculate the shear, moment, rotation, and displacement at points along a pin-pin beam. Concentrated moments, uniformly distributed forces, and linearly distributed forces are each modeled using an analogous set of concentrated forces. The approximate solution is compared with an exact solution for each of the example problems presented. The second program calculates shear, moment, rotation, and deflection for single span beams with various loading conditions and boundary conditions using the exact representation for the discontinuity functions.

7.1 Review of Discontinuity Functions

Discontinuity functions are frequently employed to describe the behavior of beams under loading. A short review of discontinuity functions is presented in this section. Gere and Timoshenko [16] provide a more comprehensive discussion of these functions in their text on mechanics of materials. Two types[2] of discontinuity functions are used to describe beam loadings: Macaulay functions and singularity functions.

Discontinuity functions are used to represent loadings which begin at a prescribed location (designated by the parameter a in the equations which follow) and continue indefinitely. By definition, Macaulay functions have the value of zero at

---

[1] In this chapter the beams obey Hooke's Law and the deflection of a beam is considered small, or $\theta \approx \tan\theta$.

[2] Both types of functions are sometimes mistakenly referred to jointly as singularity functions [16].

199

all positions to the left of the beginning position. The general form of a Macaulay function is:

$$F_n(x) = <x-a>^n = \begin{cases} 0 & \text{when } x \leq a \\ (x-a)^n & \text{when } x \geq a \end{cases}$$

$$n = 0, 1, 2, 3, \ldots$$

The use of the special < > brackets indicates that the bracketed quantity is to be replaced as specified based on the location x along the axis. It should be noted that Macaulay functions do not have singularities and are limited to zero and positive values for their exponents. This function can be integrated as follows:

$$\int_{-\infty}^{x} F_n \, dx = \int_{-\infty}^{x} <x-a>^n \, dx$$

$$= \frac{1}{n+1} <x-a>^{n+1}$$

$$n = 0, 1, 2, 3, \ldots$$

Singularity functions, in contrast to Macaulay functions, have singularities by definition. These functions are characterized by the occurrence of negative exponents for the bracketed quantities, or

$$F_n(x) = <x-a>^n = \begin{cases} 0 & \text{when } x \neq a \\ \pm\infty & \text{when } x = a \end{cases}$$

$$n = -1, -2, -3, \ldots$$

For singularity functions the bracketed quantity is replaced by terms different from those in Macaulay functions. For the case when $x = a$ the function is singular. This is caused by the fact that for negative exponents $x - a$ will produce division by zero when $x = a$. Singularity functions can be integrated as follows:

$$\int_{-\infty}^{x} F_n \, dx = \int_{-\infty}^{x} <x-a>^n \, dx$$

$$= <x-a>^{n+1}$$

$$n = -1, -2, -3, \ldots$$

Singularity functions are useful for describing special loading cases such as concentrated forces and concentrated moments (both types of loadings are demonstrated in the example problems). It should be observed that the process of integration causes singularity functions to convert to Macaulay functions.

7.2 Application of Discontinuity Functions to Beams

The discontinuity functions presented in the previous section can be used to craft a single equation representing the loading of a beam. From this load equation, w, four additional equations can be determined which represent the shear, moment, rotation, and deflection as a function of the location x along the beam, or

$$w = f(x)$$

$$V = \int w\, dx + C_v$$

$$M = \int V\, dx + C_m$$

$$EI\theta = \int M\, dx + C_\theta$$

$$EI\delta = \int\int M\, dx\, dx + C_\delta$$

where V is the shear, M is the moment, θ is the rotation, δ is the deflection, E is the modulus of elasticity, and I is the moment of inertia. The constants of integration are denoted by C_v, C_m, C_θ, and C_δ. The use of this set of relationships is dependent on the establishment of a sign convention. Figure 7.1 summarizes the sign conventions (i.e. positive definitions) which are commonly used in beam analysis.

7.3 Approximate Beam Analysis Using Superposition

The solution of complex problems described by linear equations can often be found by breaking the problem into a set of separate, less difficult problems. Typically the goal is to break the complex problem into a set of problems for which each individual solution is readily obtained. The solution of these individual problems is then combined to produce the solution to the original problem. This technique is commonly referred to as the *principle of superposition*.

In this section the shear, moment, rotation, and deflection for simply supported beams will be determined by replacing the beam loads with an appropriate set of concentrated loads. For example, a uniformly distributed load will be replaced by a set of concentrated loads placed at regular spacing which produces an approximately equivalent static loading. This approach leads to a simplified solution to even the most complex loading patterns and demonstrates the use of superposition for finding approximate solutions.

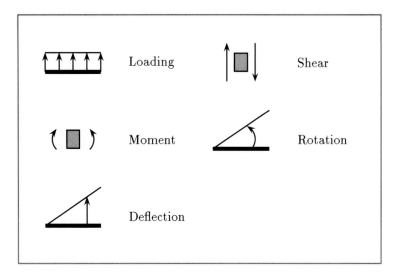

Figure 7.1. Sign Conventions

7.3.1 Equations for a Concentrated Load

The method to be employed for solving a simply supported beam requires the repetitious application of the solution for a concentrated load. Figure 7.2 shows the configuration for a pin-pin beam subjected to a positive concentrated load located within the span of the beam.

Figures 7.3 and 7.4 show the relationships for replacing uniformly and linearly distributed load segments with a single concentrated force. The distributed load for some short length is replaced with a single load located at the midpoint of the segment with length Δx. It should be obvious that for the case of the linearly distributed load that this approach does not provide a statically equivalent moment since the concentrated load is not placed at the centroid of the segment. However, increasing the discretization of the replacement loads will minimize the effects produced by this approximation.

The necessary beam relationships for a concentrated load can be easily written using discontinuity functions. Additionally, the constants of integration can be determined from statics. These equations can be summarized as

$$w = P<x-a>^{-1}$$

$$V = P<x-a>^{0} + C_v$$

$$M = P<x-a>^{1} + C_v x + C_m$$

7.3. Approximate Beam Analysis Using Superposition

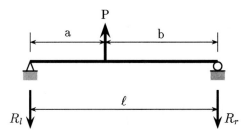

Figure 7.2. Beam With Single Concentrated Force

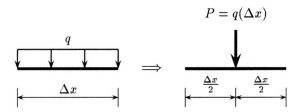

Figure 7.3. Uniformly Distributed Load Beam Segment

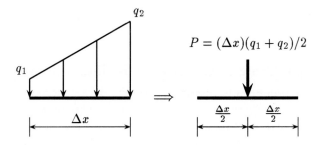

Figure 7.4. Linearly Distributed Load Beam Segment

$$EI\theta = \frac{P}{2}<x-a>^2 + \frac{C_v}{2}x^2 + C_m x + C_\theta$$

$$EI\delta = \frac{P}{6}<x-a>^3 + \frac{C_v}{6}x^3 + \frac{C_m}{2}x^2 + C_\theta x + C_\delta$$

where the boundary conditions are used to determine the constants of integration, or

$$V(x=0) = R_l \;\;\Rightarrow\;\; C_v = -\frac{Pb}{\ell}$$

$$M(x=0) = 0 \;\;\Rightarrow\;\; C_m = 0$$

$$\delta(x=0) = 0 \;\;\Rightarrow\;\; C_\delta = 0$$

$$\delta(x=\ell) = 0 \;\;\Rightarrow\;\; C_\theta = \frac{Pb}{6\ell}\left(\ell^2 - b^2\right)$$

A set of concentrated forces can be used to describe the majority of loading conditions which a beam can be subjected. A uniformly distributed load can be replaced by dividing the section subjected to the distributed load into a number of segments, each of length Δx, and placing an equivalent concentrated force for each segment at the midpoint of the segment. A similiar approach can be used for linearly distributed loads. One more loading which must be considered is a concentrated moment. The concentrated moment will be modeled using two forces acting as a couple. A *small* distance will be used to separate the forces of the couple. This approximation for a concentrated moment should be carefully studied since it artificially produces a large value of shear in the region between the two forces forming the couple.

The application of the methods just described will be demonstrated using two example problems. For each example the program calculates the approximate and true values for the shear, moment, rotation, and deflection. The difference between the approximate and true results are compared to determine the accuracy of the method. The reader should carefully study the examples to understand the nuances of this method.

7.3.2 Example Using A Uniformly Distributed Load

First, the case of a uniformly distributed load[3] will be investigated. Figure 7.5 provides the necessary description of this problem. This problem was selected to assist with the verification of the program. The loading is symmetric and therefore, the resulting relationships for shear, moment, rotation, and deflection should be symmetric. (When developing computer programs it is always prudent to use problems with intuitive solutions for verfication.) The discontinuity equations for this

---

[3] A linearly distributed loading example is included in the program but not discussed.

7.3. Approximate Beam Analysis Using Superposition

example will be used to determine the true values for shear, moment, rotation, and deflection. The approximate results will be compared with the results produced by these equations to determine the accuracy. The relationships for shear, moment, rotation, and deflection are:

$$w = -10 <x-0>^0$$

$$V = -10 <x-0>^1 + C_v$$

$$M = -\frac{10}{2} <x-0>^2 + C_v x + C_m$$

$$EI\theta = -\frac{10}{6} <x-0>^3 + \frac{C_v}{2} x^2 + C_m x + C_\theta$$

$$EI\delta = -\frac{10}{24} <x-0>^4 + \frac{C_v}{6} x^3 + \frac{C_m}{2} x^2 + C_\theta x + C_\delta$$

where the boundary conditions can be used to determine the constants of integration, or

$$V(x=0) = R_l \Rightarrow C_v = 600$$

$$M(x=0) = 0 \Rightarrow C_m = 0$$

$$\delta(x=0) = 0 \Rightarrow C_\delta = 0$$

$$\delta(x=\ell) = 0 \Rightarrow C_\theta = -720000$$

The results from the analysis for a uniformly distributed load are shown in Figures 7.6 through 7.9. Notice, as expected, the results for each relationship are symmetric. In the figures the solid line represents the true results produced from the discontinuity functions. Symbols are used to represent the results from the approximate solution. Additionally, the difference between the approximate values and the true values is plotted along the beam. For this analysis the uniformly distributed load was replaced by 120 concentrated loads along the beam. The error curve for the shear diagram has no significance since it is basically numerical noise produced by calculating values very close to zero. The same is true for the error curve for the moment diagram. This is evident from the fact that for both the shear diagram and the moment diagram the error magnitude is sixteen orders of magnitude less than the values for shear and moment. This characteristic changes for the rotation and deflection diagrams due to the inclusion of the EI term which has a magnitude of order nine. This causes the difference between the error magnitude for rotation and the value of rotation to be reduced to an order of four. The same is true for the deflection diagram. It is interesting to note that the error terms are

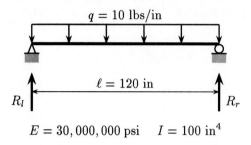

Figure 7.5. Example Using Uniformly Distributed Load

symmetric about the beam centerline for both the rotation and deflection, implying that they are the results of the numerical process and not simply numerical noise. Inspection of the results for the true and the approximate values indicates that the approximate solution is satisfactory for this loading.

7.3. Approximate Beam Analysis Using Superposition

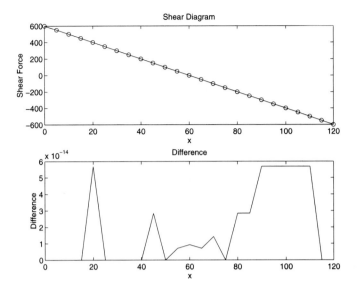

Figure 7.6. Shear Diagram for Uniformly Distributed Load

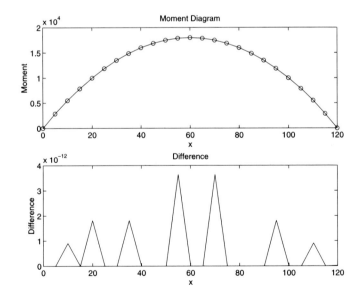

Figure 7.7. Moment Diagram for Uniformly Distributed Load

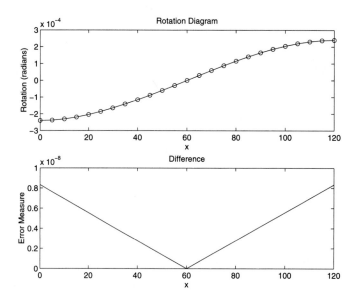

Figure 7.8. Rotation Diagram for Uniformly Distributed Load

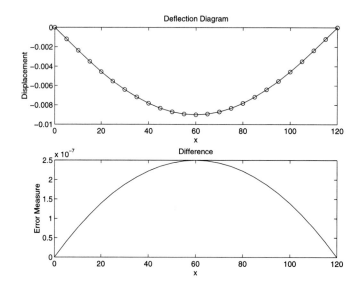

Figure 7.9. Deflection Diagram for Uniformly Distributed Load

7.3. Approximate Beam Analysis Using Superposition

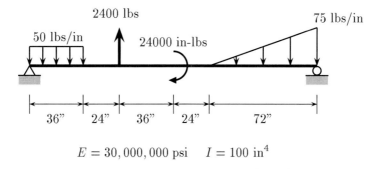

$$E = 30,000,000 \text{ psi} \quad I = 100 \text{ in}^4$$

Figure 7.10. Example Beam with Multiple Loads

7.3.3 Example with Multiple Loadings

A more complicated example using several types of loadings is now examined. Figure 7.10 provides the necessary description of this problem. Once again discontinuity equations will be used to determine the true values for shear, moment, rotation, and deflection. The approximate results will be compared with the results produced by these equations to determine the accuracy. The relationships for shear, moment, rotation, and deflection are:

$$w = -50 <x-0>^0 + 50 <x-36>^0 +$$
$$2400 <x-60>^{-1} + 24000 <x-96>^{-2} -$$
$$\frac{75}{72} <x-120>^1$$

$$V = -50 <x-0>^1 + 50 <x-36>^1 +$$
$$2400 <x-60>^0 + 24000 <x-96>^{-1} -$$
$$\frac{75}{2(72)} <x-120>^2 + C_v$$

$$M = -\frac{50}{2} <x-0>^2 + \frac{50}{2} <x-36>^2 +$$
$$2400 <x-60>^1 + 24000 <x-96>^0 -$$
$$\frac{75}{6(72)} <x-120>^3 + C_v x + C_m$$

$$EI\theta = -\frac{50}{6}<x-0>^3 + \frac{50}{6}<x-36>^3 +$$

$$\frac{2400}{2}<x-60>^2 + 24000<x-96>^1 -$$

$$\frac{75}{24(72)}<x-120>^4 + \frac{C_v}{2}x^2 + C_m x + C_\theta$$

$$EI\delta = -\frac{50}{24}<x-0>^4 + \frac{50}{24}<x-36>^4 +$$

$$\frac{2400}{6}<x-60>^3 + \frac{24000}{2}<x-96>^2 -$$

$$\frac{75}{120(72)}<x-120>^5 + \frac{C_v}{6}x^3 +$$

$$\frac{C_m}{2}x^2 + C_\theta x + C_\delta$$

where the boundary conditions can be used to determine the constants of integration, or

$$V(x=0) = R_l \Rightarrow C_v = 193.75$$

$$M(x=0) = 0 \Rightarrow C_m = 0$$

$$\delta(x=0) = 0 \Rightarrow C_\delta = 0$$

$$\delta(x=\ell) = 0 \Rightarrow C_\theta = 1848855$$

The results from this analysis are shown in Figures 7.11 through 7.14. Recall that the solid line represents the true results produced from the discontinuity functions and symbols are used to represent the results from the approximate solution. Additionally, the difference between the approximate values and the true values is also plotted. For this analysis the uniformly distributed load was replaced by 72 concentrated loads, the linearly distributed load was replaced by 144 concentrated loads, and the concentrated moment was replaced using a couple with the loads separated by one inch. Figure 7.11 shows the problem caused by the use of a couple for the moment. The shear diagram, as plotted, appears to be constant (which it isn't) because the couple distorts the y-axis of the plot. Compare this diagram to the true shear diagram presented later in this chapter in Figure 7.18. The moment diagram suffers from a similar, but not as disruptive, characteristic in the region where the couple is applied. The rotation and deflection diagrams, however, produce acceptable results.

The program implemented to approximate the solution of a beam using only concentrated loads is summarized in Table 7.1. The output for the example with

7.3. Approximate Beam Analysis Using Superposition

multiple loads is provided in Section 7.3.5. It is very important to note that the accuracy of the solution can typically be increased by using a greater number of replacement loads for distributed loads. The choice of the number of points at which to calculate the shear, moment, rotation, and deflection has no relationship with the accuracy of the results.

Table 7.1. Description of Code in Example **super**

Routine	Line	Operation
super		script file to execute program.
	23-25	select one of the example problems.
	27-30	call function **setup** to define the input values.
	32-37	initialize variables and arrays.
	41-92	loop to process each type of loading in problem.
	43-51	set various variables for instantaneous and distributed loads.
	52-56	calculate the load value for instantaneous and uniformly distributed loads.
	60-90	loop to determine contribution from each replacement load.
	61	uncomment this statement on slow computers to indicate that work is being performed.
	64	calculate the location of the load.
	65-68	calculate the value of the load for a linearly distributed load.
	70	compute left reaction by summing moments about the right support and keep a running total of these contributions (which will be produce the value of the left reaction).
	71	compute right reaction by summing moments about the left support and keep a running total of these contributions (which will be produce the value of the right reaction).
		continued on next page

continued from previous page

Routine	Line	Operation
	75-76	calculate two of the constants of integration.
	80-88	loop to determine values of shear, moment, rotation, and deflection along the beam.
	81	uncomment this statement on slow computers to indicate that work is being performed.
	82-87	include the contribution from the concentrated load if the point being considered is to the right of the load application point.
	93-95	fetch the true values for the shear, moment, rotation, and deflection along the beam by calling function **true**.
	99-104	determine the difference between the true and approximate values.
	107-110	divide the rotation and deflection values by EI.
	113-152	output the results.
	155-165	plot the shear diagram and its error diagram.
	168-178	plot the moment diagram and its error diagram.
	181-191	plot the rotation diagram and its error diagram.
	194-203	plot the deflection diagram and its error diagram.
setup		function which specifies the values used to describe the problem.
	46-53	input for multiple loading example.
	58-63	input for uniformly distributed loading example.
	68-73	input for linearly distributed loading example (not discussed in text).
true		function which calculates the true values of shear, moment, rotation, and deflection along the beam using discontinuity functions.

continued on next page

7.3. Approximate Beam Analysis Using Superposition

Routine	Line	Operation
		continued from previous page
	34-71	true values for multiple loading example.
	77-85	true values for uniformly distributed loading example.
	91-99	true values for linearly distributed loading example (not discussed in text).

7.3.4 Exercises

1. Modify the example program to solve beam problems with different boundary conditions (such as fixed-free beams).

2. Add the capability to the example program to specify parabolic distributed loads.

3. Add the capability for the user to specify a set of beam locations at which to calculate the deflections.

4. Add the capability to incorporate changes in inertia properties along the length of the beam.

5. Is it possible to use the example program to model distributed moment loads? If yes, how would it be done? If no, why is it not possible?

6. Explore the effects of varying the distance between forces for the couples used to model concentrated moments.

7. Develop an analogous program which models the loads using a distributed constant load rather than a concentrated load. What is the difference in results from the two approaches?

8. Using the case of a uniformly distributed load over the entire beam explore the relationship between the replacement load subdivision (i.e. increase the discretization of the concentrated replacement loads) and the resulting accuracy of the displacements. As the number of subdivisions are increased is there a point at which the accuracy gets worse? Or, is more subdivision always better?

9. Modify the example program to allow the maximum difference between the calculated and the true displacements to be specified. The program will have to be enhanced to include the ability to increase the loading subdivision level until the error tolerance is satisfied.

10. Modify the program to place the concentrated load at the centroid of the linearly distributed beam segments and evaluate the affects of this change in relation to placing the load at the midpoint. (Reference Figure 7.4.)

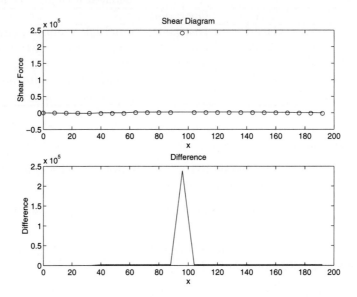

Figure 7.11. Shear Diagram for Multiple Loadings

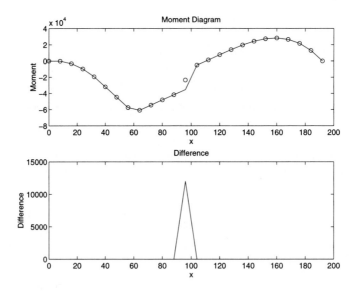

Figure 7.12. Moment Diagram for Multiple Loadings

7.3. Approximate Beam Analysis Using Superposition

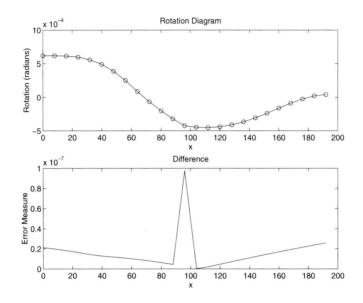

Figure 7.13. Rotation Diagram for Multiple Loadings

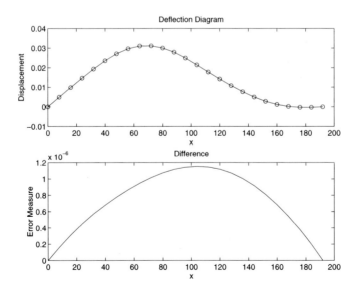

Figure 7.14. Deflection Diagram for Multiple Loadings

7.3.5 Program Output and Code

Output from Example super

```
Approximations for Simply Supported Beam
----------------------------------------

Beam length:                 192
Modulus of elasticity:       3e+07
Moment of inertia:           100
Number of deflection points: 25
Number of idealized loadings: 5

Load number 1
  Starting position
    x:                0
    load magnitude: -50
  Ending position
    x:                36
    load magnitude: -50
  # subdivisions:    72

Load number 2
  Starting position
    x:                60
    load magnitude: 2400
  Ending position
    x:                60
    load magnitude: 2400
  # subdivisions:    1

Load number 3
  Starting position
    x:                95.95
    load magnitude: 240000
  Ending position
    x:                95.95
    load magnitude: 240000
  # subdivisions:    1

Load number 4
  Starting position
    x:                96.05
    load magnitude: -240000
  Ending position
```

7.3. Approximate Beam Analysis Using Superposition 217

```
        x:               96.05
     load magnitude: -240000
   # subdivisions:    1

Load number 5
  Starting position
        x:              120
     load magnitude: 0
  Ending position
        x:              192
     load magnitude: -75
   # subdivisions:    144

Left pin reaction:
  Approximate: 193.758
  True:        193.75

Right pin reaction:
  Approximate: 1906.24
  True:        1906.25
```

Script File super

```
 1: % Example: super
 2: % ~~~~~~~~~~~~~~
 3: % This example calculates the shear, moment,
 4: % rotation, and deflection for a simply
 5: % supported beam using an approximate technique
 6: % based on superposition.  Various loads are
 7: % replaced with a system of concentrated loads
 8: % which approximate the original loading
 9: % configuration.  The results from these
10: % calculations are compared with the true
11: % results.  The beam equations used in this
12: % example were developed using discontinuity
13: % functions.
14: %
15: % All the input to this program is defined
16: % in function setup.
17: %
18: % User m functions required:
19: %    setup, true, genprint
20: %-------------------------------------------
```

```
21:
22: clear;
23: Problem=1;        % Multi-load example
24: %Problem=2;       % Uniformly distributed load
25: %Problem=3;       % Linearly distributed load
26:
27: %...Parameter definitions for specific problem
28: [L,E,I,no_defl_pts,start_x,end_x, ...
29:    start_val,end_val,no_sub_loads,no_loads]= ...
30:    setup(Problem);
31:
32: %...Initialize
33: right_react=0; left_react=0;
34: EI=E*I; dx=L/(no_defl_pts-1); x=[0:dx:L];
35: V=zeros(1,no_defl_pts); M=zeros(1,no_defl_pts);
36: EItheta=zeros(1,no_defl_pts);
37: EIdelta=zeros(1,no_defl_pts);
38:
39: %...Loop on each loading
40: %......................
41: for i=1:no_loads
42:    %...x spacing and slope of substitute loads
43:    if no_sub_loads(i) == 1
44:       %...not a distributed load
45:       dx=1; slope=0; loop_limit=1; shift_x=0;
46:    else
47:       dx=(end_x(i)-start_x(i))/no_sub_loads(i);
48:       slope=(end_val(i)-start_val(i))/ ...
49:             (end_x(i)-start_x(i));
50:       loop_limit=no_sub_loads(i); shift_x=dx/2;
51:    end
52:    %...Instantaneous load value
53:    if start_val(i) == end_val(i)
54:       %...Point loads & uniformly distributed
55:       load_val=start_val(i)*dx;
56:    end
57:
58:    %...Substitute load loop
59:    %......................
60:    for j = 1:loop_limit
61:       % fprintf('.');     % Let you know work is
62:                           %    being done
63:       %...Instantaneous location
64:       load_x=start_x(i)+dx*(j-1)+shift_x;
```

7.3. Approximate Beam Analysis Using Superposition

```
65:      if start_val(i) ~= end_val(i)
66:        %...linearly distributed load
67:        load_val=0.5*slope*dx^2*(2*j-1);
68:      end
69:      %...Superpose reaction contributions
70:      left_react=left_react-load_val*(L-load_x)/L;
71:      right_react=right_react-load_val*load_x/L;
72:
73:      %...Loop on each x along entire beam
74:      %................................
75:      bl6=L*6; b=L-load_x; Cv=-load_val*b/L;
76:      Ctheta=(load_val*b/6)*(L-b^2/L);
77:      V=V+Cv; M=M+Cv*x;
78:      EItheta=EItheta+Cv/2*x.^2+Ctheta;
79:      EIdelta=EIdelta+Cv/6*x.^3+Ctheta*x;
80:      for k = 1:no_defl_pts
81:        % fprintf('.');
82:        if (x(k)-load_x) > 0
83:          xd=x(k)-load_x; V(k)=V(k)+load_val;
84:          M(k)=M(k)+load_val*xd;
85:          EItheta(k)=EItheta(k)+load_val/2*xd^2;
86:          EIdelta(k)=EIdelta(k)+load_val/6*xd^3;
87:        end
88:      end
89:    end
90: end
91:
92: %...Calculate the true values
93: [V_true,M_true,EItheta_true,EIdelta_true, ...
94:    left_react_true,right_react_true]= ...
95:    true(Problem,L,x);
96:
97: %...Calculate the difference between the
98: %...true and approximate solutions
99: V_diff=V_true-V; M_diff=M_true-M;
100: theta_diff=EItheta_true-EItheta;
101: delta_diff=EIdelta_true-EIdelta;
102: V_diff=abs(V_diff); M_diff=abs(M_diff);
103: theta_diff=abs(theta_diff);
104: delta_diff=abs(delta_diff);
105:
106: %...Divide by EI
107: theta=EItheta/EI; theta_true=EItheta_true/EI;
108: delta=EIdelta/EI; delta_true=EIdelta_true/EI;
```

```
109:  theta_diff=theta_diff/EI;
110:  delta_diff=delta_diff/EI;
111:
112:  %...Output
113:  fprintf('\n\nApproximations for Simply ');
114:  fprintf('Supported Beam\n');
115:  fprintf(    '---------------------------');
116:  fprintf('--------------\n');
117:  fprintf( ...
118:     '\nBeam length:               %g', L);
119:  fprintf( ...
120:     '\nModulus of elasticity:     %g', E);
121:  fprintf( ...
122:     '\nMoment of inertia:         %g', I);
123:  fprintf( ...
124:     '\nNumber of deflection points: %g', ...
125:     no_defl_pts);
126:  fprintf( ...
127:     '\nNumber of idealized loadings: %g', ...
128:     no_loads);
129:  for i=1:no_loads
130:     fprintf('\n\nLoad number %g', i);
131:     fprintf( '\n  Starting position');
132:     fprintf( '\n     x:              %g', ...
133:        start_x(i));
134:     fprintf( '\n     load magnitude: %g', ...
135:        start_val(i));
136:     fprintf( '\n  Ending position');
137:     fprintf( '\n     x:              %g', ...
138:        end_x(i));
139:     fprintf( '\n     load magnitude: %g', ...
140:        end_val(i));
141:     fprintf( '\n  # subdivisions:    %g', ...
142:        no_sub_loads(i));
143:  end
144:  fprintf('\n\nLeft pin reaction:');
145:  fprintf( '\n  Approximate: %g',left_react);
146:  fprintf( '\n  True:        %g', ...
147:     left_react_true);
148:  fprintf('\n\nRight pin reaction:');
149:  fprintf( '\n  Approximate: %g',right_react);
150:  fprintf( '\n  True:        %g', ...
151:     right_react_true);
152:  fprintf('\n\n')
```

7.3. Approximate Beam Analysis Using Superposition

```
153:
154: %...Plot the shear results
155: clf;
156: subplot(2,1,1);
157:   plot(x,V,'o',x,V_true,'-');
158:   xlabel('x'); ylabel('Shear Force');
159:   title('Shear Diagram');
160: subplot(2,1,2);
161:   plot(x,V_diff)
162:   xlabel('x'); ylabel('Difference');
163:   title('Difference'); drawnow;
164: % genprint('shear');
165: disp('Press a key to continue'); pause;
166:
167: %...Plot the moment results
168: clf;
169: subplot(2,1,1);
170:   plot(x,M,'o',x,M_true,'-');
171:   xlabel('x'); ylabel('Moment');
172:   title('Moment Diagram');
173: subplot(2,1,2);
174:   plot(x,M_diff)
175:   xlabel('x'); ylabel('Difference');
176:   title('Difference'); drawnow;
177: % genprint('moment');
178: disp('Press a key to continue'); pause;
179:
180: %...Plot the rotation results
181: clf;
182: subplot(2,1,1);
183:   plot(x,theta,'o',x,theta_true,'-');
184:   xlabel('x'); ylabel('Rotation (radians)');
185:   title('Rotation Diagram');
186: subplot(2,1,2);
187:   plot(x,theta_diff)
188:   xlabel('x'); ylabel('Error Measure');
189:   title('Difference'); drawnow;
190: % genprint('theta');
191: disp('Press a key to continue'); pause;
192:
193: %...Plot the displacement results
194: clf;
195: subplot(2,1,1);
196:   plot(x,delta,'o',x,delta_true,'-');
```

```
197:    xlabel('x'); ylabel('Displacement');
198:    title('Deflection Diagram');
199: subplot(2,1,2);
200:    plot(x,delta_diff)
201:    xlabel('x'); ylabel('Error Measure');
202:    title('Difference'); drawnow;
203: % genprint('displ');
```

Function setup

```
 1: function [L,E,I,no_defl_pts, ...
 2:     start_x,end_x,start_val,end_val, ...
 3:     no_sub_loads,no_loads]=setup(Problem)
 4: %
 5: % [L,E,I,no_defl_pts,start_x,end_x, ...
 6: %    start_val,end_val,no_sub_loads, ...
 7: %    no_loads]=setup(Problem)
 8: %~~~~~~~~~~~~~~~~~~~~~~~~~~~~~~~~~~~~~~~
 9: % This function returns the parameters
10: % necessary to describe the problem
11: % indicated by Problem.
12: %
13: % Problem      - integer specification
14: %                indicating which problem
15: % L            - beam length
16: % EI           - product of E and I
17: % no_defl_pts  - number of equally spaced
18: %                points along beam at which
19: %                to calculate results
20: % start_x      - vector of starting locations
21: %                for loadings
22: % end_x        - vector of ending locations
23: %                for loadings
24: % start_val    - vector of load values which
25: %                correspond to locations in
26: %                start_x
27: % end_val      - vector of load values which
28: %                correspond to locations in
29: %                end_x
30: % no_sub_loads- vector containing the number
31: %                of substitute concentrated
32: %                loads to use.
```

7.3. Approximate Beam Analysis Using Superposition

```
33: % no_loads     - number of loading cases
34: %                required for analysis
35: %
36: % E            - modulus of elasticity
37: % I            - moment of inertia
38: %
39: % User m functions called:  none.
40: %-------------------------------------------
41:
42: %................................
43: %...Example with multiple loads
44: %................................
45: if Problem == 1
46:   no_defl_pts=25;
47:   L=192; E=30000000; I=100; EI=E*I;
48:   start_x=[ 0 60 95.95 96.05 120];
49:   end_x=  [36 60 95.95 96.05 192];
50:   start_val=[-50 2400 240000 -240000    0];
51:   end_val=  [-50 2400 240000 -240000  -75];
52:   no_sub_loads=[72 1 1 1 144];
53:   no_loads=length(no_sub_loads);
54: %................................................
55: %...Uniformly distributed load over entire beam
56: %................................................
57: elseif Problem == 2
58:   no_defl_pts=25;
59:   L=120; E=30000000; I=100; EI=E*I;
60:   start_x=[ 0]; end_x=  [120];
61:   start_val=[-10]; end_val=  [-10];
62:   no_sub_loads=[120];
63:   no_loads=length(no_sub_loads);
64: %................................................
65: %...Linearly distributed load over entire beam
66: %................................................
67: elseif Problem == 3
68:   no_defl_pts=11;
69:   L=100; E=30000000; I=100; EI=E*I;
70:   start_x=[ 0]; end_x=  [100];
71:   start_val=[-0]; end_val=  [-50];
72:   no_sub_loads=[100];
73:   no_loads=length(no_sub_loads);
74: end
```

Function true

```
1:  function [V,M,EItheta,EIdelta,left_react, ...
2:    right_react]=true(Problem,L,x)
3:  %
4:  % [V,M,EItheta,EIdelta,left_react, ...
5:  %   right_react]=true(Problem,L,x)
6:  % ~~~~~~~~~~~~~~~~~~~~~~~~~~~~~~~~~~
7:  % This function returns the true values for
8:  % V, M, EItheta, EIdelta, and reactions for
9:  % the problem indicated by the parameter
10: % Problem.
11: %
12: % Problem     - integer specification
13: %                indicating which problem
14: % L           - beam length
15: % x           - vector of equally spaced
16: %                points along beam at which
17: %                to calculate results
18: %
19: % V           - vector of shear values at x
20: % M           - vector of moment values at x
21: % EItheta     - vector of rotations at x
22: % EIdelta     - vector of deflections at x
23: % left_react  - reaction at left pin
24: % right_react - reaction at right pin
25: %
26: % User m functions called:  none.
27: %-------------------------------------------
28:
29: [no_defl_pts]=length(x);
30: %............................
31: %...Example with multiple loads
32: %............................
33: if Problem == 1
34:   left_react=193.75; right_react=1906.25;
35:   Ctheta=1848855; Cv=left_react;
36:   V(1)=left_react; M(1)=0;
37:   EItheta(1)=Ctheta; EIdelta(1)=0;
38:   for i=2:no_defl_pts
39:     xi=x(i);
40:     Vi=-50*xi+Cv; Mi=-25*xi^2+Cv*xi;
41:     EIthetai=-(50/6)*xi^3+(Cv/2)*xi^2+Ctheta;
```

7.3. Approximate Beam Analysis Using Superposition

```
42:       EIdeltai=-(50/24)*xi^4+ ...
43:               (Cv/6)*xi^3+Ctheta*xi;
44:     if (xi-36) > 0
45:       Vi=Vi+50*xi; Mi=Mi+(50/2)*(xi-36)^2;
46:       EIthetai=EIthetai+(50/6)*(xi-36)^3;
47:       EIdeltai=EIdeltai+(50/24)*(xi-36)^4;
48:     end
49:     if (xi-60) > 0
50:       Vi=Vi+2400; Mi=Mi+2400*(xi-60);
51:       EIthetai=EIthetai+(2400/2)*(xi-60)^2;
52:       EIdeltai=EIdeltai+(2400/6)*(xi-60)^3;
53:     end
54:     if (xi-96) > 0
55:       Mi=Mi+24000;
56:       EIthetai=EIthetai+(24000)*(xi-96);
57:       EIdeltai=EIdeltai+(24000/2)*(xi-96)^2;
58:     end
59:     if (xi-120) > 0
60:       Vi=Vi-(75/(72*2))*(xi-120)^2;
61:       Mi=Mi-(75/(72*6))*(xi-120)^3;
62:       EIthetai=EIthetai- ...
63:               (75/(72*24))*(xi-120)^4;
64:       EIdeltai=EIdeltai- ...
65:               (75/(72*120))*(xi-120)^5;
66:     end
67:     V(i)=Vi; M(i)=Mi;
68:     EItheta(i)=EIthetai; EIdelta(i)=EIdeltai;
69:   end
70:   V(no_defl_pts)=-right_react;
71:   M(no_defl_pts)=0; EIdelta(no_defl_pts)=0;
72:
73: %.........................................
74: %...Constant distributed load over entire beam
75: %.........................................
76: elseif Problem == 2
77:   left_react=600; right_react=600;
78:   Cv=left_react; Ctheta=-720000;
79:   V=-10*x+Cv; M=-(10/2)*x.^2+Cv*x;
80:   EItheta=-(10/6)*x.^3+(Cv/2)*x.^2+Ctheta;
81:   EIdelta=-(10/24)*x.^4+(Cv/6)*x.^3+Ctheta*x;
82:   V(no_defl_pts)=-right_react;
83:   M(no_defl_pts)=0;
84:   EItheta(no_defl_pts)=-Ctheta;
85:   EIdelta(no_defl_pts)=0;
```

```
 86: 
 87: %.........................................
 88: %...Linearly distributed load over entire beam
 89: %.........................................
 90: elseif Problem == 3
 91:    left_react=1/6*1/2*L^2;
 92:    right_react=1/3*1/2*L^2;
 93:    Cv=left_react; Ctheta=-(7/360)*1/2*L^4;
 94:    V=-0.5/2*x.^2+Cv;
 95:    M=-0.5/6*x.^3+Cv*x;
 96:    EItheta=-0.5/24*x.^4+Cv/2*x.^2+Ctheta;
 97:    EIdelta=-0.5/120*x.^5+Cv/6*x.^3+Ctheta*x;
 98:    V(no_defl_pts)=-right_react;
 99:    M(no_defl_pts)=0; EIdelta(no_defl_pts)=0;
100: end
```

7.4 Analysis of Single Span Beams Using Discontinuity Functions

In the previous section the shear, moment, rotation, and deflection of a single span beam were determined by superposing results obtained from repeatly applying a discontinuity function for a concentrated load to approximate a more complicated loading. In this section discontinuity functions are utilized to calculate these values exactly for single span beams subjected to various loading and boundary conditions. A loading classification table (i.e. discontinuity functions) and a boundary classification table (i.e. support types) are employed to describe the characteristics for a particular problem. The ability to analyze additional loadings and boundary conditions can be incorporated by expanding the tables and the corresponding computer code.

7.4.1 Program Development

Figure 7.15 summarizes five types of loading which will be incorporated into this program. The general representation for the corresponding discontinuity function is provided in the table and is shown in the positive sense for loads on the interior of the beam. Figure 7.16 shows six possible boundary conditions for a single span beam. Of the six support combinations, the fixed-fixed, fixed-pin, and pin-fixed are statically indeterminate. The program developed in this section will utilize the classification for loadings and boundary conditions (only two of the six are implemented) presented in these two figures. The program will automatically superpose the contributions from multiple load specifications and generate a plot of the relationships for shear, moment, rotation, and deflection. The sequence of steps which the program follows can be summarized as follows:

7.4. Analysis of Single Span Beams Using Discontinuity Functions

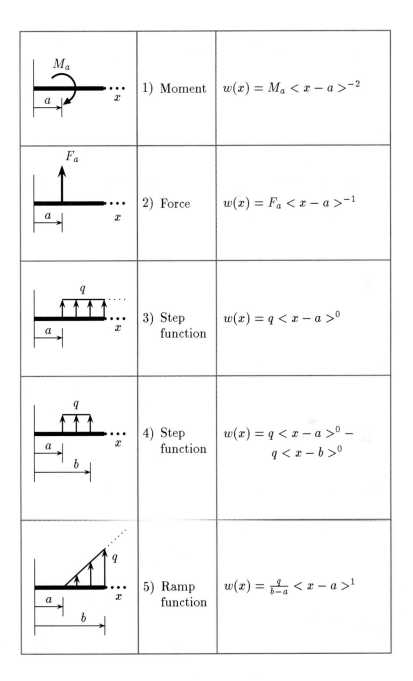

Figure 7.15. Discontinuity Functions

1. Define the beam geometry, Young's modulus, load type and corresponding loading parameters, support type, and number of locations along the beam at which the discontinuity functions will be evaluated.

2. Check for valid specification of load type and support type.

3. Isolate and identify boundary loads (those loads shown in Figure 7.16) from the input loading specifications. These loads must be identified in order to calculate the values for the coefficients of integration. The sign for the boundary moments M_1 and M_2 are shown in their positive directions in Figure 7.16.

4. Classify the characteristics of the discontinuity functions for each load. As an example consider the first load type – a concentrated moment (or load type number 1).

$$w(x) = M_a <x-a>^{-2}$$
$$V(x) = M_a <x-a>^{-1}$$
$$M(x) = M_a <x-a>^{0}$$
$$EI\theta(x) = M_a <x-a>^{1}$$
$$EI\delta(x) = M_a/2 <x-a>^{2}$$

The discontinuity coefficient is zero for w and V since the value for a discontinuity function with a negative exponent is zero by definition. The coefficient for M and θ is M_a. The coefficient for δ is $M_a/2$. Also, this discontinuity expression can be described using a single term (versus the two terms required for load type number 4). The exponent for the w function is -2 and the value for a has been specified as input to the program. This provides all the necessary information to automatically evaluate the contribution from each discontinuity function.

5. Evaluate the discontinuity functions and accumulate the total contributions from all loads for $x = \ell$. (Exclude the contribution from boundary loads.) This information is necessary to calculate the coefficients of integration which are dependent on the right end conditions.

6. Calculate the constants of integration for the discontinuity functions. The solution of simultaneous equations will be necessary for some support types.

7. Evaluate the discontinuity functions at each position x along the beam and combine with the boundary conditions to determine the values for w, V, M, θ, and δ.

The boundary conditions for each type of beam must be considered in order to develop relationships which can be used to calculate the constants of integration. The set of discontinuity functions which describe the pin-pin support type are

7.4. Analysis of Single Span Beams Using Discontinuity Functions

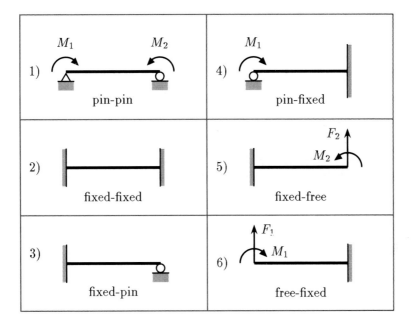

Figure 7.16. Beam Boundary Conditions

$$w(x) = w(x)_{\text{loads}}$$

$$V(x) = V(x)_{\text{loads}} + C_v$$

$$M(x) = M(x)_{\text{loads}} + C_v x + C_m$$

$$EI\theta(x) = \theta(x)_{\text{loads}} + \frac{C_v x^2}{2} + C_m x + C_\theta$$

$$EI\delta(x) = \delta(x)_{\text{loads}} + \frac{C_v x^3}{6} + \frac{C_m x^2}{2} + C_\theta x + C_\delta$$

The "loads" subscript is used to identify the total contribution from the discontinuity functions for all of the applied loads (excluding loads on the boundaries) at any position x. The constants of integration are denoted by C_v, C_m, C_θ, and C_δ and can be found from the boundary conditions, or

$$M(x = 0) = M_1 \quad \Rightarrow \quad C_m = M_1$$

$$\delta(x = 0) = 0 \quad \Rightarrow \quad C_\delta = 0$$

$$M(x = \ell) = M_2 \quad \Rightarrow \quad C_v = \frac{1}{\ell}(M_2 - M(x = \ell)_{\text{loads}} - M_1)$$

$$\delta(x = \ell) = 0 \quad \Rightarrow \quad C_\theta = -\frac{1}{\ell}\left(\delta(x = \ell)_{\text{loads}} + \frac{C_v \ell^3}{6} + \frac{M_1 \ell^2}{2}\right)$$

where $x = \ell$ coincides with the right end of the beam.

The set of discontinuity functions which describe the fixed-fixed support type are

$$w(x) = w(x)_{\text{loads}}$$

$$V(x) = V(x)_{\text{loads}} + C_v$$

$$M(x) = M(x)_{\text{loads}} + C_v x + C_m$$

$$EI\theta(x) = \theta(x)_{\text{loads}} + \frac{C_v x^2}{2} + C_m x + C_\theta$$

$$EI\delta(x) = \delta(x)_{\text{loads}} + \frac{C_v x^3}{6} + \frac{C_m x^2}{2} + C_\theta x + C_\delta$$

and the constants of integration are

$$\theta(x = 0) = 0 \quad \Rightarrow \quad C_\theta = 0$$

$$\delta(x = 0) = 0 \quad \Rightarrow \quad C_\delta = 0$$

$$\theta(x = \ell) = 0 \quad \Rightarrow \quad \frac{C_v \ell^2}{2} + C_m \ell + \theta(x = \ell)_{\text{loads}} = 0$$

$$\delta(x = \ell) = 0 \quad \Rightarrow \quad \frac{C_v \ell^3}{6} + \frac{C_m \ell^2}{2} + \delta(x = \ell)_{\text{loads}} = 0$$

It is obvious that simultaneous equations must be solved to determine constants C_v and C_m.

The program just summarized is straight forward but requires careful attention to the multitude of details necessary to implement the many combinations of loads and supports. A description of the program is contained in Table 7.2 and the actual code is included in Section 7.4.4.

7.4. Analysis of Single Span Beams Using Discontinuity Functions

Table 7.2. Description of Code in Example **singspan**

Routine	Line	Operation
singspan		script file to execute program.
	38	select one of the example problems.
	39-69	input definitions for example problems.
	72-84	make sure the support and load types specified have been implemented.
	90-147	handle loads applied at the boundaries and define the discontinuity function coefficients and parameters for each load type.
	149-153	evaluate the discontinuity functions at $x = \ell$ using function **seval**. These values will be necessary for determining the constants of integration.
	155-158	determine the constants of integration using function **constant**.
	164-177	loop to evaluate the discontinuity functions at the specified locations along the beam.
	165-167	evaluate the discontinuity functions at $x = x_i$ using function **seval**.
	169-176	include the contributions from the constants of integration.
	179-225	output the results.
	228-256	plot the results.
seval		function which evaluates the discontinuity functions for a specified value of x.
	43-77	loop through each loading.
	44-76	loop to process each term of a discontinuity function. For example, load type four in Figure 7.15 requires two terms in the equation for its loading function.
		continued on next page

continued from previous page

Routine	Line	Operation
	45	compute the value of the bracketed expression.
	46	only perform calculations when the bracketed expression is greater or equal to zero.
	47-54	contribution to loading function.
	55-62	contribution to shear function.
	63-68	contribution to moment function.
	69-71	contribution to rotation function.
	72-74	contribution to deflection function.
constant		function which calculates the values for the constants of integration.
	35-39	values for pin-pin supports.
	40-46	values for fixed-fixed supports.
	43-46	calculate the constants of integration for the fixed-fixed supports by solving a set of simultaneous equations.

7.4. Analysis of Single Span Beams Using Discontinuity Functions

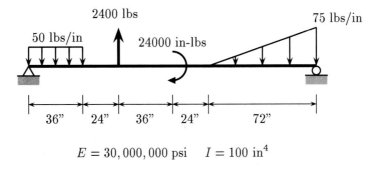

$$E = 30,000,000 \text{ psi} \quad I = 100 \text{ in}^4$$

Figure 7.17. Example Beam with Multiple Loads

7.4.2 Example with Multiple Loadings

The multiple loading example presented in the previous section is employed to illustrate the use of the methods developed for analyzing single span beams. A description of this example is shown in Figure 7.17. The shear, moment, rotation, and deflection plots generated from the analysis for 48 positions along the beam are depicted in Figures 7.18 and 7.19.

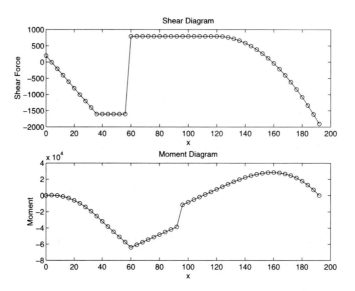

Figure 7.18. Shear and Moment Diagram

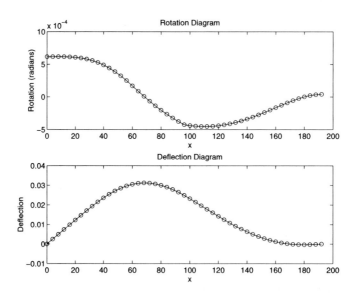

Figure 7.19. Rotation and Deflection Diagram

7.4. Analysis of Single Span Beams Using Discontinuity Functions

7.4.3 Exercises

1. Implement the capability to analyze beams with boundary conditions 3 and 4 shown in Figure 7.16.

2. Implement the capability to analyze beams with boundary conditions 5 and 6 shown in Figure 7.16.

3. Implement the capability to analyze beams with the loading type 6 shown in Figure 7.20.

4. Implement the capability to analyze beams with the loading type 7 shown in Figure 7.20.

7.4.4 Program Output and Code

Output from Example singspan

```
Single Span Beam Analysis
Using Discontinuity Functions
-----------------------------

Problem Specifications:
  Beam length:        192
  E:                  3e+07
  I:                  100
  # of beam segments: 48
  # of loads:         4
  Type of support:    1 (pin-pin)

Summary for each load:
  Load 1
    Magnitude:   -50
    Starting x:  0
    Ending x:    36
    Load type:   4 (step function, finite)
  Load 2
    Magnitude:   2400
    Starting x:  60
    Ending x:    60
    Load type:   2 (concentrated force)
  Load 3
    Magnitude:   24000
    Starting x:  96
```

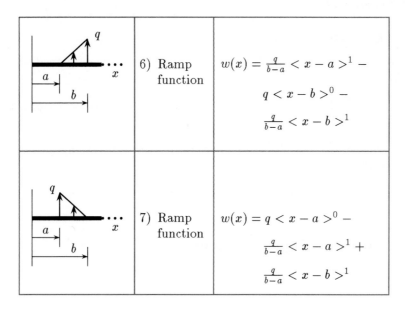

Figure 7.20. Additional Discontinuity Functions

```
    Ending x:    96
    Load type:   1 (concentrated moment)
  Load 4
    Magnitude:   -75
    Starting x:  120
    Ending x:    192
    Load type:   5 (ramp function, continuous)

Boundary results:
  Left end of beam
    Shear:       193.75
    Moment:      0
    Rotation:    0.000616285
    Deflection:  0
  Right end of beam
    Shear:       -1906.25
    Moment:      0
    Rotation:    4.0285e-05
    Deflection:  0
```

7.4. Analysis of Single Span Beams Using Discontinuity Functions

Script File singspan

```
 1: % Example: singspan
 2: % ~~~~~~~~~~~~~~~~~
 3: % This example calculates the shear, moment,
 4: % rotation, and deflection for a single span
 5: % beams using discontinuity functions.
 6: %
 7: % Date is defined in the declaration statements
 8: % below, where:
 9: %
10: % Blen       - beam length
11: % Emod       - modulus of elasticity
12: % Binert     - beam moment of inertia
13: % No_segs    - number of locations to evaluate.
14: %              Calculations will be for every
15: %              (Blen/No_segs) location
16: % S_type     - type of beam supports
17: %              =1, pin-pin
18: %              =2, fixed-fixed
19: % Load_type  - vector containing type of load
20: %              for each loading
21: %              =1, concentrated moment
22: %              =2, concentrated force
23: %              =3, step function, continuous
24: %              =4, step function, finite
25: %              =5, ramp function, continuous
26: % Load       - vector containing load value
27: %              for each loading
28: % Start_x    - vector containing starting
29: %              location for each loading
30: % End_x      - vector containing ending
31: %              location for each loading
32: %
33: % User m functions required:
34: %     seval, constant, genprint
35: %-----------------------------------------------
36:
37: clear;
38: Problem=3;
39: if Problem == 1
40:   %..pin-pin example
41:   Blen=192; Emod=29e6; Binert=394;
```

```
42:    No_segs=16; S_type=1;
43:    Load_type=[2         4            ];
44:    Load=     [-12000   -416.66667];
45:    Start_x=  [132       60           ];
46:    End_x=    [132       132          ];
47: elseif Problem == 2
48:    %...fixed-fixed example
49:    Blen=300; Emod=29e6; Binert=60;
50:    No_segs=15; S_type=2;
51:    Load_type=[3]; Load=[-20];
52:    Start_x=[150]; End_x=[300];
53: elseif Problem == 3
54:    %...Multiple loading example, pin-pin
55:    Blen=192; Emod=30e6; Binert=100;
56:    No_segs=48; S_type=1;
57:    Load_type=[4      2        1       5  ];
58:    Load=     [-50   2400    24000   -75];
59:    Start_x=  [0      60       96     120];
60:    End_x=    [36     60       96     192];
61: elseif Problem == 4
62:    %...Boundary moments only, pin-pin
63:    Blen=10; Emod=30e6; Binert=100;
64:    No_segs=10; S_type=1;
65:    Load_type=[1  1 ];
66:    Load=     [10 10];
67:    Start_x=  [0  10];
68:    End_x=    [0  10];
69: end
70: No_loads=length(Load_type);
71:
72: %...Check valid ranges
73: if S_type < 1 | S_type > 2
74:    fprintf('\n\nInvalid support type error: ');
75:    fprintf('%g \n\n',S_type);
76:    error('Program aborted');
77: end
78: for i=1:No_loads
79:    if Load_type(i) < 1 | Load_type(i) > 5
80:       fprintf('\n\nInvalid load type error: ');
81:       fprintf('%g \n\n',Load_type(i));
82:       error('Program aborted');
83:    end
84: end
85:
```

7.4. Analysis of Single Span Beams Using Discontinuity Functions

```
86:  %...Define the coefficients of the
87:  %...discontinuity functions and handle
88:  %...the loads which may be applied
89:  %...on the boundaries
90:  M1=0; M2=0;
91:  Wl=0; Vl=0; Ml=0; Thetal=0; Deltal=0;
92:  for j=1:No_loads
93:    %...Handle each load type
94:    if Load_type(j) == 1
95:      %...Concentrated moment
96:      if Start_x(j) == 0
97:        %...Left boundary load
98:        M1=Load(j); Nterms(j)=0;
99:      elseif Start_x(j) == Blen
100:       %...Right boundary load
101:       M2=Load(j); Nterms(j)=0;
102:     else
103:       W(j,1)=0; V(j,1)=0; M(j,1)=Load(j);
104:       Theta(j,1)=Load(j);
105:       Delta(j,1)=Load(j)/2;
106:       Nterms(j)=1; Power(j,1)=-2;
107:       A(j,1)=Start_x(j);
108:     end
109:   elseif Load_type(j) == 2
110:     W(j,1)=0; V(j,1)=Load(j);
111:     M(j,1)=Load(j);
112:     Theta(j,1)=Load(j)/2;
113:     Delta(j,1)=Load(j)/6;
114:     Nterms(j)=1; Power(j,1)=-1;
115:     A(j,1)=Start_x(j);
116:   elseif Load_type(j) == 3
117:     %...Step function, continuous
118:     W(j,1)=Load(j); V(j,1)=Load(j);
119:     M(j,1)=Load(j)/2;
120:     Theta(j,1)=Load(j)/6;
121:     Delta(j,1)=Load(j)/24;
122:     Nterms(j)=1; Power(j,1)=0;
123:     A(j,1)=Start_x(j);
124:   elseif Load_type(j) == 4
125:     %...Step function, finite
126:     W(j,1)=Load(j); V(j,1)=Load(j);
127:     M(j,1)=Load(j)/2;
128:     Theta(j,1)=Load(j)/6;
129:     Delta(j,1)=Load(j)/24;
```

```
130:      W(j,2)=-Load(j); V(j,2)=-Load(j);
131:      M(j,2)=-Load(j)/2;
132:      Theta(j,2)=-Load(j)/6;
133:      Delta(j,2)=-Load(j)/24;
134:      Nterms(j)=2; Power(j,1)=0; Power(j,2)=0;
135:      A(j,1)=Start_x(j); A(j,2)=End_x(j);
136:    elseif Load_type(j) == 5
137:      %...Ramp function
138:      %... (continuous and increasing)
139:      dx=End_x(j)-Start_x(j);
140:      W(j,1)=Load(j)/dx; V(j,1)=Load(j)/(2*dx);
141:      M(j,1)=Load(j)/(6*dx);
142:      Theta(j,1)=Load(j)/(24*dx);
143:      Delta(j,1)=Load(j)/(120*dx);
144:      Nterms(j)=1; Power(j,1)=1;
145:      A(j,1)=Start_x(j);
146:    end
147: end
148:
149: %...Evaluate the discontinuity functions at
150: %...x=Beam_length
151: [W1,V1,M1,Theta1,Delta1]= ...
152:    seval(Blen,Nterms,No_loads,Power,A, ...
153:         W,V,M,Theta,Delta);
154:
155: %...Determine constants of integration
156: [Cv,Cm,Ctheta,Cdelta]= ...
157:    constant(M1,M2,Blen,S_type, ...
158:            V1,M1,Theta1,Delta1);
159:
160: %...Evaluate the discontinuity functions
161: %...along the beam
162: N1=No_segs+1; x=linspace(0,Blen,N1);
163: WX=0; VX=0; MX=0; ThetaX=0; DeltaX=0;
164: for j=1:N1
165:    [WX,VX,MX,ThetaX,DeltaX]= ...
166:       seval(x(j),Nterms,No_loads,Power,A, ...
167:            W,V,M,Theta,Delta);
168:    %...Combine with boundary contributions
169:    Wx(j)=WX;
170:    Vx(j)=VX+Cv;
171:    Mx(j)=MX+Cv*x(j)+Cm;
172:    Thetax(j)=(ThetaX+Cv*x(j)^2/2+ ...
173:             Cm*x(j)+Ctheta)/(Emod*Binert);
```

7.4. Analysis of Single Span Beams Using Discontinuity Functions

```
174:      Deltax(j)=(DeltaX+Cv*x(j)^3/6+ ...
175:              Cm*x(j)^2/2+Ctheta*x(j)+ ...
176:              Cdelta)/(Emod*Binert);
177: end
178:
179: %...Output results
180: fprintf('\n\n   Single Span Beam Analysis');
181: fprintf('\nUsing Discontinuity Functions');
182: fprintf('\n----------------------------');
183: fprintf('\n\nProblem Specifications:');
184: fprintf('\n  Beam length:         %g',Blen);
185: fprintf('\n  E:                   %g',Emod);
186: fprintf('\n  I:                   %g',Binert);
187: fprintf('\n  # of beam segments:  %g',No_segs);
188: fprintf('\n  # of loads:          %g',No_loads);
189: fprintf('\n  Type of support:     %g',S_type);
190: if S_type == 1
191:    fprintf(' (pin-pin)');
192: elseif S_type == 2
193:    fprintf(' (fixed fixed)');
194: end
195: fprintf('\n\nSummary for each load:');
196: for i=1:No_loads
197:    fprintf('\n  Load %g',i);
198:    fprintf('\n    Magnitude:  %g',Load(i));
199:    fprintf('\n    Starting x: %g',Start_x(i));
200:    fprintf('\n    Ending x:   %g',End_x(i));
201:    fprintf('\n    Load type:  %g',Load_type(i));
202:    if Load_type(i) == 1
203:       fprintf(' (concentrated moment)');
204:    elseif Load_type(i) == 2
205:       fprintf(' (concentrated force)');
206:    elseif Load_type(i) == 3
207:       fprintf(' (step function, continuous)');
208:    elseif Load_type(i) == 4
209:       fprintf(' (step function, finite)');
210:    elseif Load_type(i) == 5
211:       fprintf(' (ramp function, continuous)');
212:    end
213: end
214: fprintf('\n\nBoundary results:');
215: fprintf('\n  Left end of beam');
216: fprintf('\n    Shear:     %g',Vx(1));
217: fprintf('\n    Moment:    %g',Mx(1));
```

```
218:  fprintf('\n    Rotation:  %g',Thetax(1));
219:  fprintf('\n    Deflection: %g',Deltax(1));
220:  fprintf('\n Right end of beam');
221:  fprintf('\n    Shear:     %g',Vx(N1));
222:  fprintf('\n    Moment:    %g',Mx(N1));
223:  fprintf('\n    Rotation:  %g',Thetax(N1));
224:  fprintf('\n    Deflection: %g',Deltax(N1));
225:  fprintf('\n\n');
226:
227:  %...Plot the results
228:  clf;
229:  plot(x,Wx,'-',x,Wx,'o');
230:    xlabel('x'); ylabel('Load');
231:    title('Load Diagram'); drawnow;
232:  % genprint('bload');
233:    disp('Press a key to continue'); pause;
234:
235:  clf;
236:  subplot(2,1,1);
237:    plot(x,Vx,'-',x,Vx,'o');
238:    xlabel('x'); ylabel('Shear Force');
239:    title('Shear Diagram');
240:  subplot(2,1,2);
241:    plot(x,Mx,'-',x,Mx,'o');
242:    xlabel('x'); ylabel('Moment');
243:    title('Moment Diagram'); drawnow;
244:  % genprint('bshrmom');
245:    disp('Press a key to continue'); pause;
246:
247:  clf;
248:  subplot(2,1,1);
249:    plot(x,Thetax,'-',x,Thetax,'o');
250:    xlabel('x'); ylabel('Rotation (radians)');
251:    title('Rotation Diagram');
252:  subplot(2,1,2);
253:    plot(x,Deltax,'-',x,Deltax,'o');
254:    xlabel('x'); ylabel('Deflection');
255:    title('Deflection Diagram'); drawnow;
256:  % genprint('brotdef');
```

7.4. Analysis of Single Span Beams Using Discontinuity Functions

Function seval

```
 1: function [Wx,Vx,Mx,Thetax,Deltax]= ...
 2:    seval(xloc,Nterms,Nloads,Power,A, ...
 3:        W,V,M,Theta,Delta);
 4: %
 5: % [Wx,Vx,Mx,Thetax,Deltax]= ...
 6: %    seval(xloc,Nterms,Nloads,Power,A, ...
 7: %        W,V,M,Theta,Delta);
 8: %~~~~~~~~~~~~~~~~~~~~~~~~~~~~~~~~~~~~~
 9: % This function evaluates the discontinuity
10: % functions at a specified value of x.
11: %
12: % xloc    - location along beam
13: % Nterms  - vector containing the number
14: %           of terms required to model
15: %           each type of loading
16: % Nloads  - number of loadings
17: % Power   - matrix of exponents for each
18: %           load term
19: % A       - matrix of a values for the
20: %           bracket part of function for
21: %           each load term
22: % W       - matrix of load coefficients
23: %           for each bracket term
24: % V       - matrix of shear coefficients
25: %           for each bracket term
26: % M       - matrix of moment coefficients
27: %           for each bracket term
28: % Theta   - matrix of rotation coefficients
29: %           for each bracket term
30: % Delta   - matrix of deflection coefficients
31: %           for each bracket term
32: %
33: % Wx      - load at specified x
34: % Vx      - shear at specified x
35: % Mx      - moment at specified x
36: % Thetax  - rotation at specified x
37: % Deltax  - deflection at specified x
38: %
39: % User m functions called:  none.
40: %-------------------------------------------
41:
```

```
42: Wx=0; Vx=0; Mx=0; Thetax=0; Deltax=0;
43: for j=1:Nloads
44:    for k=1:Nterms(j)
45:       bracket=xloc-A(j,k);
46:       if bracket >= 0
47:          if Power(j,k) >= 0
48:             %...W-x
49:             tmp=1;
50:             if Power(j,k) > 0
51:                tmp=bracket^Power(j,k);
52:             end
53:             Wx=Wx+W(j,k)*tmp;
54:          end
55:          if Power(j,k)+1 >= 0
56:             %...V-x
57:             tmp=1;
58:             if Power(j,k)+1 > 0
59:                tmp=bracket^(Power(j,k)+1);
60:             end
61:             Vx=Vx+V(j,k)*tmp;
62:          end
63:          %...M-x
64:          tmp=1;
65:          if Power(j,k)+2 > 0
66:             tmp=bracket^(Power(j,k)+2);
67:          end
68:          Mx=Mx+M(j,k)*tmp;
69:          %...Theta-x
70:          Thetax=Thetax+ ...
71:                 Theta(j,k)*bracket^(Power(j,k)+3);
72:          %...Delta-x
73:          Deltax=Deltax+ ...
74:                 Delta(j,k)*bracket^(Power(j,k)+4);
75:       end
76:    end
77: end
```

Function constant

```
1: function [Cv,Cm,Ctheta,Cdelta]= ...
2:    constant(M1,M2,Blen,S_type, ...
3:             V1,M1,Theta1,Delta1);
```

7.4. Analysis of Single Span Beams Using Discontinuity Functions

```
%
% [Cv,Cm,Ctheta,Cdelta]= ...
%    constant(M1,M2,Blen,S_type, ...
%             V1,M1,Theta1,Delta1);
% ~~~~~~~~~~~~~~~~~~~~~~~~~~~~~~~~~
% This function determines the constants
% of integration for the beam.
%
% M1      - applied moment at left boundary
% M2      - applied moment at right boundary
% Blen    - length of beam
% S_type  - type of support system
%           =1, pin-pin
%           =2, fixed-fixed
% V1      - shear value for discontinuity
%           function evaluated at x=Blen
% M1      - moment value for discontinuity
%           function evaluated at x=Blen
% Theta1  - rotation value for discontinuity
%           function evaluated at x=Blen
% Delta1  - deflection value for discontinuity
%           function evaluated at x=Blen
%
% Cv      - shear constant of integration
% Cm      - moment constant of integration
% Ctheta  - rotation constant of integration
% Cdelta  - deflection constant of integration
%
% User m functions called: none.
%-------------------------------------------------

if S_type == 1
  %...pin-pin
  Cv=(M2-M1-M1)/Blen; Cm=M1; Cdelta=0;
  Ctheta=-(Delta1+Cv*Blen^3/6+ ...
          Cm*Blen^2/2)/Blen;
elseif S_type == 2
  %...fixed-fixed
  Ctheta=0; Cdelta=0;
  a=[Blen^2/2   Blen     ; ...
     Blen^3/6   Blen^2/2];
  b=[-Theta1; -Delta1];
  x=a\b; Cv=x(1); Cm=x(2);
end
```

Chapter 8

Additional Topics

This chapter includes problems related to several additional topics in mechanics of materials. The flange bolt problem developed in Chapter 2 is reformulated for a general shear connection. The analysis of a beam column is considered and the moment and deflection in a cantilever beam column is studied. A general solution to the problem of beams constructed from several materials is employed to investigate the behavior of a thermostat constructed using a bimetalic beam. Finally, a generally stiffness solution is developed to analyze planar trusses.

8.1 Stresses in an Eccentric Shear Connection

Structural components are often bolted, or riveted, together to create a load bearing connection. Figure 8.1 depicts a typical arrangement for an eccentrically loaded shear connection. The bolts used for the connection must be designed and sized to carry the shear stresses induced by the loads being transferred between structural members. A common assumption used in the design of the bolts is that all the shear is transferred by the bolts (the surfaces of the structural elements in contact are frictionless). A second assumption which is employed is that the shearing deformations are proportional to and normal to the radius measured from the centroid of the bolts. In this section a short program is presented which can be used to determine the shear stress in each of the bolts in an eccentrically loaded shear connection.

8.1.1 Development of Stress Relationships

The total shear stress in each bolt can be determined by summing the contribution of shear stress from the force and the moment acting on the connector, or

$$\tau = \tau_{\text{force}} + \tau_{\text{moment}}$$

The total moment acting at the centroid of the bolts is

$$M = M_o + F_x \bar{y} - F_y \bar{x}$$

8.1. Stresses in an Eccentric Shear Connection

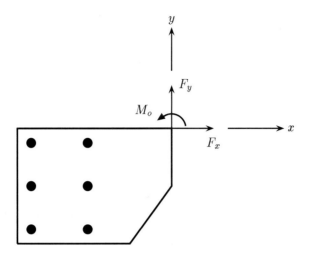

Figure 8.1. Eccentrically Load Shear Connection

where positive directions for the loads correspond to the directions shown in Figure 8.1. The centroid of the bolts can be found using

$$\bar{x} = \frac{\sum x_i A_i}{\sum A_i} \qquad \bar{y} = \frac{\sum y_i A_i}{\sum A_i}$$

where A_i is the area of a bolt, (x_i, y_i) is the coordinate of the center of the bolt, and the summation is taken for the n bolts in the connector. It is convenient to translate all the coordinates to a centroidal coordinate system, or

$$\hat{x} = x - \bar{x} \qquad \hat{y} = y - \bar{y}$$

As previously stated, the stresses developed in the bolts can be separated into the contribution from the forces and the contribution from the moment. The total shear stress contribution developed from the two force components is

$$(\tau_{\text{force}})_x = \frac{F_x}{A} \qquad (\tau_{\text{force}})_y = \frac{F_y}{A}$$

where the total area of the bolts can be found from

$$A = \sum_{i=1}^{n} A_i$$

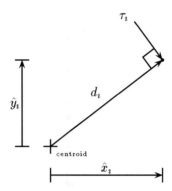

Figure 8.2. Stress Due to Moment on Single Bolt

To determine the shear stress in a bolt developed from the moment acting at the centroid two assumptions are utilized: a) the shear stress is proportional to the distance from the centroid, and b) the shear stress acts in a direction normal to the radius arm d as shown in Figure 8.2. The stress proportionality relationship produces

$$\frac{\tau_i}{d_i} = \frac{\tau_{i+1}}{d_{i+1}} = \frac{\tau_{i+2}}{d_{i+2}} = \cdots$$

where

$$d_i^2 = \hat{x}_i^2 + \hat{y}_i^2$$

This relationship can be rewritten to yield

$$\begin{aligned}
\tau_1 &= \left(\frac{d_1}{d_1}\right)\tau_1 = \tau_1 \\
\tau_2 &= \left(\frac{d_2}{d_1}\right)\tau_1 \\
\tau_3 &= \left(\frac{d_3}{d_1}\right)\tau_1 \\
&\vdots \quad \vdots \\
\tau_i &= \left(\frac{d_i}{d_1}\right)\tau_1
\end{aligned}$$

The moment acting at the centroid must be generate an equivalent amount of shear

8.1. Stresses in an Eccentric Shear Connection

stress in the bolts to produce equilibrium, or

$$M = \sum_{i=1}^{n} f_i d_i$$

where f_i is the force resultant for the shear stress created by the moment and d_i is the moment arm. Recalling that the shear stress is related to the force by

$$(\tau_{\text{moment}})_i = \frac{f_i}{A_i}$$

and substituting for f_i yields

$$M = \sum_{i=1}^{n} (\tau_i A_i d_i)$$

$$= \sum_{i=1}^{n} \left[\left(\frac{d_i}{d_1} \right) \tau_1 A_i d_i \right]$$

$$= \left(\frac{\tau_1}{d_1} \right) \sum_{i=1}^{n} (d_i^2 A_i)$$

This equation can be rewritten for τ_i, or[1]

$$\tau_i = \frac{M d_1}{\sum d_i^2 A_i}$$

The components of stress can be determined from geometry using Figure 8.2 which produces

$$(\tau_x)_i = \left(\frac{\hat{y}_i}{d_i} \right) \tau_i \qquad (\tau_y)_i = \left(\frac{\hat{x}_i}{d_i} \right) \tau_i$$

Substituting this relationship into the equation for τ_i and rearranging terms yields

$$(\tau_x)_i = \frac{M \hat{y}_i}{\sum d_i^2 A_i}$$

$$(\tau_y)_i = \frac{M \hat{x}_i}{\sum d_i^2 A_i}$$

Therefore the two components of stress on each bolt due to both the applied forces and the applied moment are

$$(\tau_x)_i = \frac{F_x}{A} - \frac{M \hat{y}_i}{\sum d_i^2 A_i}$$

$$(\tau_y)_i = \frac{F_y}{A} + \frac{M \hat{x}_i}{\sum d_i^2 A_i}$$

---

[1] It should be noted that this form is analogous to $\tau = Tr/J$.

8.1.2 Program to Calculate the Bolt Stresses

Program **shrcon** was written to solve for the shearing stresses in the bolts of an ecccentric shear connection. The output from an example dataset, along with the computer code, is included in Section 8.1.3. A description of the code is summarized in Table 8.1. The program also generates a plot showing the location of the bolts and the location of the load point.

Routine	Line	Operation
shrcon		script file to execute program.
	33	select one of the example problems.
	34-49	define the input parameters.
	53-55	calculate area of each bolt if bolt diameter is specified.
	58-60	determine the centroid for the bolts.
	63	shift coordinates to centroidal coordinate system.
	66	calculate the total moment acting at the centroid.
	69-70	calculate the shear stress in each bolt.
	72-94	output the results.
	97-105	plot of the bolt locations.

Table 8.1. Description of Code in Example **shrcon**

8.1.3 Program Output and Code

Output from Example shrcon

```
Stresses In Eccentric Shear Connection
--------------------------------------

Applied Loads:
  Fx:  8
  Fy: -6
  M:   0

Total moment at centroid: -66

Bolt/Rivet #: 1
  Diameter:   1
  Area:       0.785398
  x position: -9
```

8.1. Stresses in an Eccentric Shear Connection

```
    y position: -6
    Stress (x): -2.30163
    Stress (y): -5.14193
    Stress:      5.63355

Bolt/Rivet #: 2
    Diameter:    1
    Area:        0.785398
    x position: -9
    y position:  0
    Stress (x):  7.39458
    Stress (y): -5.14193
    Stress:      9.00663

Bolt/Rivet #: 3
    Diameter:    1
    Area:        0.785398
    x position: -5
    y position: -6
    Stress (x): -2.30163
    Stress (y):  1.32221
    Stress:      2.65438

Bolt/Rivet #: 4
    Diameter:    1
    Area:        0.785398
    x position: -5
    y position:  0
    Stress (x):  7.39458
    Stress (y):  1.32221
    Stress:      7.51186
```

Script File shrcon

```
 1: % Example: shrcon
 2: % ~~~~~~~~~~~~~~~
 3: % This example determines the shear stresses
 4: % in the bolts/rivets of an eccentric shear
 5: % connection.
 6: %
 7: % Data is defined in the declaration statements
 8: % below, where:
 9: %
```

```
10: % Fx              - applied force in x
11: % Fy              - applied force in y
12: % M               - applied moment
13: % Diameter        - diameters of bolt/rivets
14: % Areai           - areas of bolt/rivets
15: % x               - x positions of bolt/rivets
16: % y               - y positions of bolt/rivets
17: %
18: % NOTES: a) The origin of the coordinate
19: %             system must be located at the
20: %             point of application of the load
21: %             vector.
22: %          b) Either the "Diameter" or "Areai"
23: %             must be specified.  If "Diameter"
24: %             is specified the area is
25: %             calculated as (pi*diameter^2/4).
26: %
27: % User m functions required:
28: %     genprint
29: %-------------------------------------------
30:
31: clear;
32: %...Input definitions
33: Problem=3;
34: if Problem == 1
35:   Fx=0; Fy=-24; M=Fy*5;
36:   Diameter =[7/8 7/8 7/8 7/8 7/8 7/8];
37:   x        =[0 0 0 4 4 4];
38:   y        =[2 5 8 2 5 8];
39: elseif Problem == 2
40:   Fx=0; Fy=-24; M=0;
41:   Areai=[0.601 0.601 0.601 0.601 0.601 0.601];
42:   x     =[-7    -7    -7    -3    -3    -3];
43:   y     =[-8    -5    -2    -8    -5    -2];
44: elseif Problem == 3
45:   Fx=8; Fy=-6; M=0;
46:   Diameter =[ 1  1  1  1];
47:   x        =[-9 -9 -5 -5];
48:   y        =[-6  0 -6  0];
49: end
50: Nbolts=length(x);
51:
52: %...Area
53: if length(Areai) == 0
```

8.1. Stresses in an Eccentric Shear Connection 253

```
54:     Areai=pi*Diameter.^2/4;
55: end
56: Area=sum(Areai);
57:
58: %...Centroid of bolt system
59: xbar=sum(x.*Areai)/Area;
60: ybar=sum(y.*Areai)/Area;
61:
62: %...Shift to centroidal axes
63: xs=x-xbar; ys=y-ybar;
64:
65: J=sum(Areai.*(xs.^2+ys.^2));
66: Mtot=M+Fx*ybar-Fy*xbar;
67:
68: %...Shear stresses
69: TauX=Fx/Area-Mtot*ys/J; TauY=Fy/Area-Mtot*xs/J;
70: Tau=sqrt(TauX.^2+TauY.^2);
71:
72: fprintf( ...
73:    '\n\nStresses In Eccentric Shear Connection');
74: fprintf( ...
75:    '\n---------------------------------------');
76: fprintf('\n\nApplied Loads:');
77: fprintf(  '\n   Fx: %g',Fx);
78: fprintf(  '\n   Fy: %g',Fy);
79: fprintf(  '\n   M:  %g',M);
80: fprintf( ...
81:    '\n\nTotal moment at centroid: %g',Mtot);
82: for i=1:Nbolts
83:    fprintf('\n\nBolt/Rivet #: %g',i);
84:    if length(Diameter) > 0
85:       fprintf('\n   Diameter:    %g', Diameter(i));
86:    end
87:    fprintf('\n   Area:        %g', Areai(i));
88:    fprintf('\n   x position:  %g', x(i));
89:    fprintf('\n   y position:  %g', y(i));
90:    fprintf('\n   Stress (x):  %g', TauX(i));
91:    fprintf('\n   Stress (y):  %g', TauY(i));
92:    fprintf('\n   Stress:      %g', Tau(i));
93: end
94: fprintf('\n');
95:
96: clf;
97: plot(x,y,'o',[0],[0],'*')
```

```
 98:     axis('equal');
 99:     title( ...
100:     'Stresses In Eccentric Shear Connection');
101:     xlabel('x'); ylabel('y');
102:     tmp=legend('o',' Bolt/Rivet', ...
103:                '*',' Load Point');
104:     axes(tmp); drawnow;
105: % genprint('shrcon');
```

8.2 Combined Axial and Flexural Loading of a Beam Column

Beams subjected to a combination of axial and flexural loads are called beam columns. A configuration for a beam cantilevered at the left end and subjected to end loads as shown in Figure 8.3 is examined in this section. The unknowns to be obtained are the resulting moment and deflection throughout the beam. The boundary conditions at the left end require

$$y(0) = 0 \qquad y'(0) = 0$$

The moment equation for the beam involves the right end deflection y_0 which must be found during the analysis. The differential equation governing our problem is

$$EIy''(x) = M(x) = -P_0(y_0 - y) + F_0(\ell - x) + M_0$$

or

$$EIy''(x) - P_0 y = -F_0 x + (M_0 + F_0\ell - P_0 y_0)$$

The general solution is represented as the sum of a homogeneous solution and a particular solution. The homogeneous solution is

$$y_h = A\sinh(kx) + B\cosh(kx) \qquad k = \sqrt{\frac{P}{EI}}$$

The particular solution has the form

$$y_p = Cx + D$$

where C and D are chosen to satisfy the differential equation. The relationship defined by

$$-P_0(Cx + D) = -F_0 x + (M_0 + F_0\ell - P_0 y_0)$$

requires

$$C = \frac{F_0}{P_0} \qquad D = -\frac{(M_0 + F_0\ell - P_0 y_0)}{P_0}$$

The left end boundary conditions are used to produce

$$y(0) = 0 = B + D \qquad y'(0) = 0 = kB + C$$

8.2. Combined Axial and Flexural Loading of a Beam Column

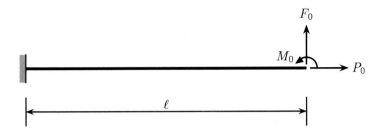

Figure 8.3. Axially Loaded Cantilever Beam

Thus, the deflection equation is

$$y(x) = C\left[x - \frac{\sinh(kx)}{k}\right] + D\left[1 - \cosh(kx)\right]$$

Recognizing that $y(\ell) = y_0$ yields

$$y_0 = \frac{F_0}{P_0}\left[\ell - \frac{\sinh(k\ell)}{k}\right] + \left[y_0 - \frac{M_0 + F_0\ell}{P_0}\right]\left[1 - \cosh(k\ell)\right]$$

and the value of y_0 is found to be

$$y_0 = \frac{\frac{M_0}{P_0}\left[\cosh(k\ell) - 1\right] + \frac{F_0}{P_0}\left[\ell\cosh(k\ell) - \frac{\sinh(k\ell)}{k}\right]}{\cosh(kl)}$$

with

$$k = \sqrt{\frac{P_0}{EI}}$$

It should be noted that changing the direction of P_0 gives

$$k = \imath\sqrt{\frac{-P_0}{EI}} \qquad \imath - \sqrt{-1}$$

with the trigonometric quantities being related by

$$\cosh(\imath\theta) = \cos\theta \qquad \sinh(\imath\theta) = \imath\sin\theta$$

Fortunately, MATLAB handles complex numbers intrinsicly and the solution using hyperbolic functions is still valid when P_0 is negative, except for the values which make $\cos\left(\ell\sqrt{\frac{-P_0}{EI}}\right)$ zero. This condition produces the Euler buckling load

$$\ell\sqrt{\frac{-P_0}{EI}} = \frac{\pi}{2}$$

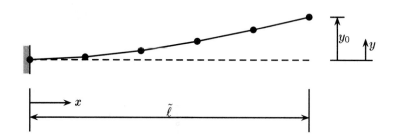

Figure 8.4. Deflected Shape of Axially Loaded Cantilever Beam

or

$$-P_0 = \frac{\pi^2 EI}{4\ell^2}$$

Taking $-P_0$ close to the buckling load will produce very large deflection values, whereas use of the exact buckling load yields physically impossible infinite deflections.

8.2.1 Program to Analyze an Axially Loaded Beam

A program was written to perform the analysis described in the preceding section and the code, along with the output for an example problem, is included in Section 8.2.3. A description of the code is summarized in Table 8.2. Figure 8.4 depicts the deflected shape of a discretized representation of the beam column. Note that the length of the beam has changed due to the axial load and is represented by $\tilde{\ell}$. Figure 8.5 includes a graph of the moment and deflection for the beam subjected to the conditions in the example problem contained in the program.

Figure 8.5 plots the deflected shape of the beam with and without the effects of the axial load. The results correspond with what might be expected from casual observation – the axial load tends to reduce the displacement for this loading configuration.

8.2.2 Exercises

1. Determine the deflected shape for the beam in the example if $P_0 = 0$. Compare this result with the the result found in the example problem by plotting both solutions on the same graph.

2. Modify the example program to create a set of curves corresponding to values of P_0 which are 90%, 94%, 98%, and 100% of the Euler buckling load. Plot the set of curves on a single graph.

8.2. Combined Axial and Flexural Loading of a Beam Column

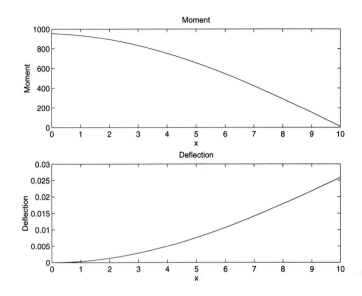

Figure 8.5. Moment and Deflection in Axially Loaded Beam

Routine	Line	Operation
axdefl		script file to execute program.
	24-32	define the input parameters.
	34-35	calculate some constants.
	36-37	calculate the deflection at the right end.
	38	calculate two constants.
	39	define set of nodes along the beam.
	41	calculate the deflections.
	42	calculate the moments.
	44-56	output the input parameters.
	57-66	output table of moments and deflections.
	68-79	plot results.

Table 8.2. Description of Code in Example **axdefl**

8.2.3 Program Output and Code

Output from Example axdefl

```
Deflection of Axially Loaded Beam
---------------------------------

Length:                  10
E:                       3e+07
Inertia:                 0.0490874
F0:                      10
M0:                      10
P0:                      -32701.9
Number of beam nodes:    101

Beam Results at Nodes:
      #       x              M              y
    --- -------------  -------------  -------------
      1  0.00000e+00    9.54880e+02    0.00000e+00
      2  1.00000e-01    9.53774e+02    3.24092e-06
      3  2.00000e-01    9.52456e+02    1.29584e-05
      4  3.00000e-01    9.50927e+02    2.91436e-05
      5  4.00000e-01    9.49186e+02    5.17860e-05
      6  5.00000e-01    9.47235e+02    8.08738e-05
      7  6.00000e-01    9.45074e+02    1.16394e-04
      8  7.00000e-01    9.42702e+02    1.58331e-04
      9  8.00000e-01    9.40121e+02    2.06670e-04

... Material deleted ...

     91  9.00000e+00    1.52000e+02    2.17993e-02
     92  9.10000e+00    1.37900e+02    2.21999e-02
     93  9.20000e+00    1.23770e+02    2.26014e-02
     94  9.30000e+00    1.09611e+02    2.30038e-02
     95  9.40000e+00    9.54290e+01    2.34069e-02
     96  9.50000e+00    8.12254e+01    2.38107e-02
     97  9.60000e+00    6.70038e+01    2.42150e-02
     98  9.70000e+00    5.27672e+01    2.46197e-02
     99  9.80000e+00    3.85190e+01    2.50249e-02
    100  9.90000e+00    2.42622e+01    2.54302e-02
    101  1.00000e+01    1.00000e+01    2.58358e-02
```

8.2. Combined Axial and Flexural Loading of a Beam Column

Script File axdefl

```
1:  % Example: axdefl
2:  % ~~~~~~~~~~~~~~~
3:  % This program determines the deflection in
4:  % a cantilever beam subjected to a transverse
5:  % load, concentrated moment, and axial load
6:  % at the free end of the beam.
7:  %
8:  % Data is defined in the declaration statements
9:  % below, where:
10: %
11: % Length   - beam length
12: % Emod     - modulus of elasticity
13: % Inertia  - beam inertia
14: % F0       - value of transverse load
15: % M0       - value of moment load
16: % P0       - value of axial load
17: % Npts     - number of beam nodes along beam for
18: %            discretization of beam
19: %
20: % User m functions required: genprint
21: %-------------------------------------------------
22:
23: clear;
24: Problem=1;
25: if Problem == 1
26:    %...Cantilever beam of circular cross section
27:    Length=10; Emod=30e6; Npts=101;
28:    F0=10; M0=10;
29:    Diameter=1; Inertia=pi*Diameter^4/64;
30:    % Axial force is 90% of the buckling load
31:    P0=-0.90*pi^2*Emod*Inertia/(2*Length)^2;
32: end
33:
34: k=sqrt(P0/(Emod*Inertia));
35: skl=sinh(k*Length); ckl=cosh(k*Length);
36: yend=M0/P0*(ckl-1)/ckl+ ...
37:      F0/P0*(Length*ckl-skl/k)/ckl;
38: c=F0/P0; d=yend-(M0+F0*Length)/P0;
39: x=linspace(0,Length,Npts)';
40: skx=sinh(k*x); ckx=cosh(k*x);
41: y=c*(x-skx/k)+d*(1-ckx);
```

```
42: moment=-Emod*Inertia*k*(c*skx+d*k*ckx);
43:
44: fprintf( ...
45:   '\n\nDeflection of Axially Loaded Beam');
46: fprintf('\n--------------------------------');
47: fprintf( ...
48:   '\n\nLength:                    %g',Length);
49: fprintf('\nE:                         %g',Emod);
50: fprintf( ...
51:   '\nInertia:                   %g',Inertia);
52: fprintf('\nF0:                        %g',F0);
53: fprintf('\nM0:                        %g',M0);
54: fprintf('\nP0:                        %g',P0);
55: fprintf( ...
56:   '\nNumber of beam nodes:      %g',Npts);
57: fprintf('\n\nBeam Results at Nodes:');
58: fprintf('\n        #        x              M       ');
59: fprintf('         y');
60: fprintf('\n      ---  ------------  ------------');
61: fprintf('  ------------');
62: for i=1:length(y)
63:   fprintf('\n  %4.0f %12.5e %12.5e %12.5e', ...
64:     i,x(i),moment(i),y(i));
65: end
66: fprintf('\n');
67:
68: clf;
69: ax=subplot(2,1,1);
70:   plot(x,moment,'-');
71:   title('Moment'); xlabel('x'); ylabel('Moment');
72:   drawnow; Rlimits=get(ax,'RenderLimits');
73: subplot(2,1,2);
74:   plot(x,y,'-');
75:   title('Deflection');
76:   xlabel('x'); ylabel('Deflection');
77:   set(gca,'XLim',[Rlimits(1),Rlimits(2)]);
78:   drawnow;
79:   %genprint('axdefl');
```

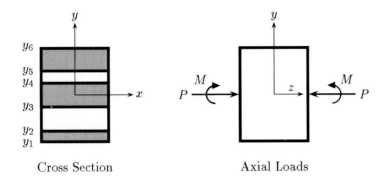

Figure 8.6. Beam Made of Several Materials

8.3 Deflection of a Nonhomogeneous Beam Subjected to Axial Loading

In this section the effects of temperature changes and applied loads in a beam of nonhomogeneous cross section are considered. The beam is assumed to have only one plane of bending. The elastic modulus E and the thermal expansion coefficient α can vary spatially. Figure 8.6 illustrates a beam composed of several rectangular layers. Initially the beam is straight and unstressed. Then the beam is subjected to a temperature change t and loads normal to the cross section which produce an axial load P and a bending moment M relative to a conveniently chosen set of reference axes.

Because the beam is nonhomogeneous, no special benefit results from placing the reference axes at the centroid. However, changing the axes could, of course, affect the moment M produced by the normal stresses on the cross section whenever $P \neq 0$.

The familiar assumption used to compute stresses in the beam is that a plane cross section remains plane after loading and temperature change. Therefore, the longitudinal strain has the form

$$\epsilon = c_1 + c_2 y = \epsilon_t + \epsilon_\ell$$

where c_1 and c_2 are to be computed using equilibrium conditions. The strain will consist of the temperature strain $\epsilon_t = \alpha t$ and the loading strain ϵ_ℓ. The accompanying stress σ will be given by

$$\sigma = (\epsilon - \epsilon_t)E = (c_1 + c_2 y - \alpha t)E$$

where E varies over the cross section. The normal stress produces an axial load P

and a moment M, or

$$P = \iint \sigma \, d(\text{area})$$
$$= c_1 \iint E \, d(\text{area}) + c_2 \iint Ey \, d(\text{area}) - t \int \alpha \, d(\text{area})$$
$$= c_1 I_1 + c_2 I_2 - t I_3$$

and

$$M = \iint y\sigma \, d(\text{area})$$
$$= c_1 \iint Ey \, d(\text{area}) + c_2 \iint Ey^2 \, d(\text{area}) - t \int \alpha y \, d(\text{area})$$
$$= c_1 I_2 + c_2 I_4 - t I_5$$

There are five integrals which account for spatial variability of E and α. When the beam consists of several parts, each having constant E and α values, the analysis for the beam simplifies to give

$$I_1 = \iint E \, d(\text{area}) = \sum_{j=1}^{n} E_j A_j$$

$$I_2 = \iint E\bar{y} \, d(\text{area}) = \sum_{j=1}^{n} E_j Q_j = \sum_{j=1}^{n} E_j A_j \bar{y}_j$$

$$I_3 = \iint \alpha \, d(\text{area}) = \sum_{j=1}^{n} \alpha_j A_j$$

$$I_4 = \iint Ey^2 \, d(\text{area}) = \sum_{j=1}^{n} E_j (I_x)_j$$

$$I_5 = \iint \alpha y \, d(\text{area}) = \sum_{j=1}^{n} \alpha_j Q_j = \sum_{j=1}^{n} \alpha_j A_j \bar{y}_j$$

where A_j is an area, Q_j is a first moment of area, I_j is a moment of inertia, $\bar{y}_j$ is a centroidal distance, subscript j denotes the j'th part, and n is the total number of parts. Constants c_1 and c_2 are obtained by solving the system

$$I_1 c_1 + I_2 c_2 = P + t I_3$$
$$I_2 c_1 + I_3 c_2 = M + t I_5$$

The normal stresses in part j can be found using

$$\sigma_j = E_j (c_1 + c_2 y - t \alpha_j)$$

8.3. Deflection of a Nonhomogeneous Beam Subjected to Axial Loading

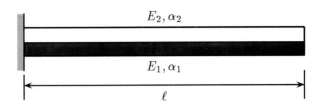

Figure 8.7. Cantilever Beam Made of Two Materials

and y can have any value compatible with the part.

Figure 8.6 depicts a nonhomogeneous beam with material interfaces parallel to the xz plane. However, the ideas presented here apply for a more general geometry. A typical example could include a circular rod embedded in a trapezoidal cross section. The analysis would still involve computation of the part areas, the first order area moments, and the inertial moments for each part which can then be used to determine c_1 and c_2.

A MATLAB function called **compbeam** was written to handle a multilayered rectangular beam as shown in Figure 8.6. This function is employed in the next section to analyze a practical example for deformation of a bimetallic strip employed in a thermostat.

8.3.1 Thermostat Constructed from a Beam of Two Materials

One method which can be used to construct a switch for a thermostat is to laminate two materials with different thermal expansion properties to make a beam as shown in Figure 8.7. A change in temperature will cause the two materials to expand, or contract, at different rates and will cause the beam to bend. A proper choice in beam materials and geometry results in the construction of a switch controlled by the variation in temperature. In this section a program is presented to analyze the affects which the various material properties have on the characteristics of the beam deflection for a cantilever beam constructed of two materials.

8.3.2 Deflection of Beam Constructed From Two Materials

The beam shown in Figure 8.7 will bend when subjected to a change in temperature, even though P and M are zero. The curvature for a beam is $y''(x)$ and is also equal to the constant c_2 discussed in the previous section. For a cantilever beam with $y(0) = 0$ and $y'(0) = 0$, the deflection is $y(x) = \frac{c_1 x^2}{2}$. The end deflection is therefore given by

$$\delta = \frac{c_2 \ell^2}{2}$$

8.3.3 Program to Analyze Beam of Two Materials

A program was written to study the influence various parameters (α, E, L, and h) have on the deflection which occurs at the right end of the beam. The output for the program is included in Section 8.3.4. A description of the code is summarized in Table 8.3. Figure 8.8 depicts the complete solution space for a combination of ratios for modulus of elasticity and coefficient of thermal expansion. Figure 8.9 shows how the modulus of elasticity ratio affects the deflection when the coefficient of thermal expansion is kept constant. Figure 8.10 is the analogous plot for the coefficient of thermal expansion when the modulus of elasticity ratio is kept constant.

Routine	Line	Operation
bimetal		script file to execute program.
	52-61	define the input parameters.
	64-76	output the input parameters.
	81-82	initialize some parameters.
	83-99	loop through set of moduli of elasticities.
	88-98	loop through set of thermal coefficients.
	90-91	use function **compbeam** to find the end deflection.
	92-97	output the results for this combination of parameters.
	102-127	plot results.
compbeam		function which determines the the flexural stress, curvature, and end deflection in a beam having having a rectangular cross section composed of several layers.

Table 8.3. Description of Code in Example **bimetal**

8.3. Deflection of a Nonhomogeneous Beam Subjected to Axial Loading

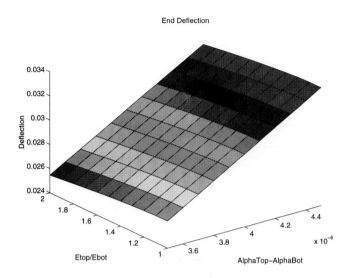

Figure 8.8. End Deflection

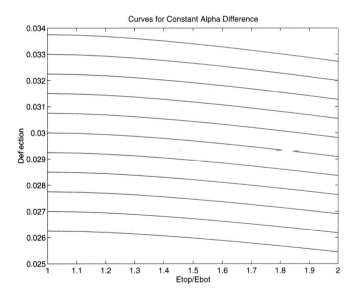

Figure 8.9. Curves for Constant Alpha Difference

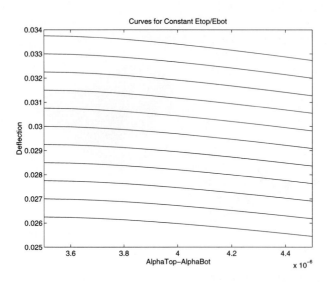

Figure 8.10. Curves for Constant Etop/Ebot

8.3.4 Program Output and Code

Output from Example bimetal

```
Beam of Two Materials
---------------------


Length:                 1
Beam width:             0.1
Beam height:            0.01
E for bottom member:    2.7e+07
Temperature change:     -200

Summary of results:

  E ratio: 1
  Etop:    2.7e+07

    Alpha difference:    3.5e-06
    AlpTop:              3e-06
    Deflection at x=L:   0.02625

    Alpha difference:    3.6e-06
    AlpTop:              3e-06
```

8.3. Deflection of a Nonhomogeneous Beam Subjected to Axial Loading

```
    Deflection at x=L:  0.027

    Alpha difference:   3.7e-06
    AlpTop:             3e-06
    Deflection at x=L:  0.02775

... Material deleted ...

  E ratio: 2
  Etop:     5.4e+07

    Alpha difference:   3.5e-06
    AlpTop:             2e-06
    Deflection at x=L:  0.0254545

... Material deleted ...

    Alpha difference:   4.4e-06
    AlpTop:             2e-06
    Deflection at x=L:  0.032

    Alpha difference:   4.5e-06
    AlpTop:             2e-06
    Deflection at x=L:  0.0327273
```

Script File bimetal

```
 1: % Example: bimetal
 2: % ~~~~~~~~~~~~~~~~
 3: % This program determines the deflection in
 4: % a cantilever beam constructed using two
 5: % materials.  The beam is subjected solely
 6: % to a change in temperature which induces
 7: % the resulting deflection.  This example
 8: % analyzes the affect of the coefficient
 9: % of thermal expansion and modulus of
10: % elasticity as it relates to the resulting
11: % deflection.
12: %
13: % Data is defined in the declaration statements
14: % below, where:
15: %
16: % Length     - beam length
```

```
17: % Bwidth     - beam width
18: % Bheight    - height of each material
19: %              (half the total beam height)
20: % Ebot       - modulus of elasticity of bottom
21: %              beam
22: % Evals      - vector of modulus of elasticity
23: %              ratios to analyze (Etop/Ebot)
24: % AlpBot     - coefficient of thermal expansion
25: %              for bottom beam
26: % Alphavals - vector of coefficient of thermal
27: %              expansion differences to analyze
28: %              AlphaTop(i)=AlpBot-Alphavals(i)
29: % DeltaTemp - change in temperature. Positive
30: %              temperature denotes a temperature
31: %              decrease.
32: % M          - applied bending moment. Positive
33: %              moment implies compression for
34: %              positive y.
35: % P          - applied axial load. Positive
36: %              load tends to cause axial
37: %              compression.
38: % Yloc       - y-location of beam interfaces,
39: %              including bottom and top
40: %              locations.
41: %
42: % where:
43: %
44: % Etop       - modulus of elasticity of top beam
45: % AlpTop     - coefficent of thermal expansion for
46: %              top beam
47: %
48: % User m functions required: compbeam, genprint
49: %-----------------------------------------------
50:
51: clear;
52: Problem=1;
53: if Problem == 1
54:    %...Cantilever bimetallic beam
55:    Length=1; DeltaTemp=-200; M=0; P=0;
56:    Bwidth=.1; Bheight=0.01;
57:    Ebot=27e6; Evals=[1:0.1:2.0];
58:    AlpBot=6.5e-6;
59:    Alphavals=[3.5e-6:0.1e-6:4.5e-6];
60:    Yloc=[0 Bheight 2*Bheight];
```

8.3. Deflection of a Nonhomogeneous Beam Subjected to Axial Loading

```
61:  end
62:  Eloop=length(Evals); Aloop=length(Alphavals);
63:
64:  fprintf('\n\nBeam of Two Materials');
65:  fprintf('\n--------------------');
66:  fprintf( ...
67:     '\n\nLength:                    %g',Length);
68:  fprintf( ...
69:     '\nBeam width:                %g',Bwidth);
70:  fprintf( ...
71:     '\nBeam height:               %g',Bheight);
72:  fprintf( ...
73:     '\nE for bottom member:       %g',Ebot);
74:  fprintf( ...
75:     '\nTemperature change:        %g',DeltaTemp);
76:  fprintf('\n');
77:
78:  fprintf('\nSummary of results:');
79:
80:  %...Loop for each elasticity ratio
81:  Etop=Evals*Ebot;
82:  AlpTop=AlpBot-Alphavals;
83:  for i=1:Eloop
84:     E=[Ebot Etop(i)];
85:     fprintf('\n\n  E ratio: %g',Evals(i));
86:     fprintf('\n  Etop:    %g',Etop(i));
87:     %...Loop for each coefficient difference
88:     for j=1:Aloop
89:        Alpha=[AlpBot AlpTop(j)];
90:        [Y,yStress,C,deltaL(i,j)]=compbeam ...
91:        (Yloc,E,Bwidth,Length,M,P,DeltaTemp,Alpha);
92:        fprintf('\n\n    Alpha difference: %g', ...
93:           Alphavals(j));
94:        fprintf('\n    AlpTop:           %g', ...
95:           AlpTop(i));
96:        fprintf('\n    Deflection at x=L: %g', ...
97:           deltaL(i,j));
98:     end
99:  end
100: fprintf('\n\n');
101:
102: clf;
103:    surf(Alphavals,Evals,deltaL);
104:    title('End Deflection');
```

```
105:    xlabel('AlphaTop-AlphaBot');
106:    ylabel('Etop/Ebot');
107:    zlabel('Deflection');
108:    drawnow;
109: % genprint('therm1a')
110:    disp('Press a key to continue'); pause;
111:
112: clf;
113:    plot(Evals,deltaL)
114:    title('Curves for Constant Alpha Difference');
115:    xlabel('Etop/Ebot');
116:    ylabel('Deflection');
117:    drawnow;
118: % genprint('therm1b')
119:    disp('Press a key to continue'); pause;
120:
121: clf;
122:    plot(Alphavals,deltaL)
123:    title('Curves for Constant Etop/Ebot');
124:    xlabel('AlphaTop-AlphaBot');
125:    ylabel('Deflection');
126:    drawnow;
127: % genprint('therm1c')
```

Function compbeam

```
 1: function [Y,S,c,yend]= ...
 2:    compbeam(y,e,w,len,m,p,t,alp)
 3: %
 4: % [Y,S,c,yend]=compbeam(y,e,w,len,m,p,t,alp)
 5: % ~~~~~~~~~~~~~~~~~~~~~~~~~~~~~~~~~~~~~~~~~
 6: % This function determines flexural stresses
 7: % and curvature in a beam having a rectangular
 8: % cross section composed of several rectangular
 9: % layers of the same width which are bonded
10: % together. Each layer can have a different
11: % elastic modulus and thermal expansion
12: % coefficient. Initially the beam is straight
13: % with zero normal stresses acting on the
14: % cross section. When axial loading, bending
15: % moment, and temperature change are applied,
16: % a plane cross section still remains plane
```

8.3. Deflection of a Nonhomogeneous Beam Subjected to Axial Loading

```
% and normal stresses in the axial direction
% occur. The axial strain has the form:
%              c(1)+c(2)*y
% where c(1) and c(2) are found by imposing
% equilibrium and strain conditions.
%
% y     - a vector defining vertical positions
%         of the bottom, the material
%         interfaces and the top of the beam
%         measured from any chosen reference
%         axes. It may be convenient, for
%         example, to take y=0 at the bottom
%         of the beam.
% e     - a vector containing values of
%         Young's modulus for each successive
%         layer.
% w,len - the beam width and length
% m     - the applied bending moment measured
%         relative to the chosen reference axes.
% p     - the axial load applied to the beam
% t     - the temperature change from the
%         initial configuration at which the
%         beam is straight and unstressed.
% alp   - a vector of thermal coefficients for
%         the various layers of the beam.
%
% Y,S   - vectors of y values and stress values
%         describing the piecewise linear
%         variation of stress as a function of
%         y. At material interfaces, the stress
%         can change discontinuously, so some
%         pairs of succesive values in vector
%         Y may be the same with corresponding
%         components of S being different.
% c     - the vector describing the axial strain
%         variation. c(1) represents a uniform
%         axial component, whereas c(2) equals
%         the beam curvature approximated by
%         y''(x).
% yend  - the transverse deflection at the
%         right end when the left end slope and
%         deflection equal zero. This value is
%         simply c(2)/2*len^2.
%
```

```
61: % where:
62: % ae    - vector of areas times elastic moduli
63: %         for successive layers.
64: % qe    - vector of first moments of area times
65: %         elastic moduli for successive layers.
66: % ie    - vector of inertial moments times
67: %         elastic moduli for successive layers.
68: %
69: % User m functions called:  none
70: %------------------------------------------------
71:
72: if nargin < 8 % No temperature change
73:   t=0; n=length(y)-1; alp=ones(1,n); end
74: y=y(:)'; e=e(:)'; n=length(y)-1; N=1:n;
75: yt=y(N+1); yb=y(N); a=w*(yt-yb);
76: q=w/2*(yt.^2-yb.^2); i=w/3*(yt.^3-yb.^3);
77: if length(e)==1 % All E the same.
78:   e=e*ones(1,n); end;
79: if length(alp)==1 % All alpha the same.
80:   alp=alp*ones(1,n); end;
81: ae=a*e'; qe=q*e'; ie=i*e'; p=p+(t*alp.*a)*e';
82: m=m+(t*alp.*q)*e'; c=[ae,qe;qe,ie]\[p;m];
83: alpe=t*alp.*e; Y=[y;y]; Y=Y(:); Y=Y(2:2*n+1);
84: E=[e;e]; E=E(:); APE=[alpe;alpe]; APE=APE(:);
85: S=c(1)*E+c(2)*Y.*E-APE; yend=c(2)/2*len^2;
```

8.4 Analysis of Pin-Connected Trusses

Plane trusses consist of a number of axially loaded members connected at node points. Trusses are used in many applications such as bridges or roof supports for wide span buildings. If the truss members are assumed to support only axial loads and member buckling is not a concern, then member loads can be obtained by solving for the horizontal and vertical displacements at each node, and then using displacements to compute loads.

Consider a typical axially loaded member which connects nodes i and j that have coodinates (x_i, y_i) and (x_j, y_j) as shown in Figure 8.11. The member k has an inclination angle θ_k and length ℓ_k such that

$$\theta_k = \tan^{-1}\left(\frac{y_j - y_i}{x_j - x_i}\right) \qquad \ell_k = \sqrt{(x_j - x_i)^2 + (y_j - y_i)^2}$$

The axial elongation of the member is

$$\Delta_k = (u_j - u_i)\cos\theta_k + (v_j - v_i)\sin\theta_k$$

8.4. Analysis of Pin-Connected Trusses

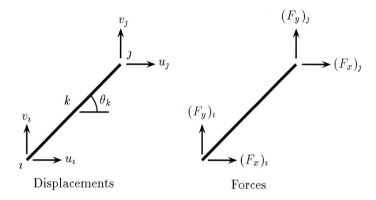

Figure 8.11. Typical Truss Member

where u represents the horizontal displacements, v represents the vertical displacements, A is the area of a truss element, and E is the modulus of elasticity for a truss element. The axial elongation gives rise to an axial force

$$P_k = \left(\frac{A_k E_k}{\ell_k}\right) \Delta_k$$

Furthermore the axial force leads to cartesian force components at the nodes expressed as

$$(F_x)_j = -(F_x)_i = P_k \cos \theta_k$$

$$(F_y)_j = -(F_y)_i = P_k \sin \theta_k$$

These formulas can be put into matrix form using c to denote $\cos \theta_k$ and s to denote $\sin \theta_k$ which yields

$$\begin{bmatrix} (F_x)_i \\ (F_y)_i \\ (F_x)_j \\ (F_y)_j \end{bmatrix} = \frac{A_k E_k}{\ell_k} \begin{bmatrix} -c \\ -s \\ c \\ s \end{bmatrix} \begin{bmatrix} -c & -s & c & s \end{bmatrix} \begin{bmatrix} u_i \\ v_i \\ u_j \\ v_j \end{bmatrix}$$

or

$$F_k = K_k U_k$$

where F_k is a matrix of force components, U_k is a matrix of displacement components and K_k is a stiffness matrix defined by

$$K_k = \left(\frac{A_k E_k}{\ell_k}\right) \begin{bmatrix} c^2 & cs & -c^2 & -cs \\ cs & s^2 & -cs & -s^2 \\ -c^2 & -cs & c^2 & cs \\ -cs & -s^2 & cs & s^2 \end{bmatrix}$$

Each member of the truss has a stiffness matrix and the contributions of all members can be assembled to give a global stiffness matrix which simultaneously describes the equilibrium of all nodes. This has the form

$$F = KU$$

where U is a vector containing all horizontal and vertical nodal displacement components and F is a vector representing the external force resultants at the nodes. The general stiffness matrix will be singular until it is modified to impose unique displacements in the truss. For example, one or several support nodes might exist with displacements specified, or some nodes could be connected to a roller forced to move tangent to a given surface as shown in Figure 8.12. Hence, two types of constraints must be specified and are characterized by:

- Type 1: displacements in a horizontal or vertical direction, and
- Type 2: the angle for the allowed roller movements.

For Type 1 conditions, either the horizontal or the vertical force equations for the appropriate nodes are replaced by the known displacement values. Type 2 conditions are similar. If a node is constrained to slide in a direction defined by angle α, then the displacement components u and v must give no displacement normal to the desired direction, or

$$u \sin \alpha - v \cos \alpha = 0$$

Similarly, the nodal force resultant along the unconstrained direction must be zero, or

$$F_x \cos \alpha + F_y \sin \alpha = 0$$

When the displacement constraints are imposed on the global equilibrium equations a solvable system is produced of the form

$$\tilde{F} = \tilde{K} U \qquad U = (\tilde{K})^{-1} \tilde{F}$$

where $\tilde{F}$ and $\tilde{K}$ are modified global force and stiffness matrices. Once the nodal displacements are known, member loads can be found by multiplying individual stiffness matrices by corresponding nodal displacements. The computation procedure can be summarized as follows

8.4. Analysis of Pin-Connected Trusses

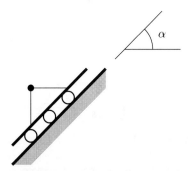

Figure 8.12. Roller Boundary Condition on Inclined Plane

1) Define the data specifying

   a) Node numbers, position coordinates, and applied nodal force components

   b) Constraint conditions on displacements at selected nodes

   c) Truss member data specifying the area, elastic modulus, and nodal indices for each member

2) Compute the stiffness matrix for each member and use these to assemble the global stiffness matrix.

3) Modify the global equilibrium conditions to impose nodal displacement constraints. Then solve the modified equations to determine all unknown nodal displacements.

4) Use the displacements to compute member loads and any unknown nodal forces where displacement constraints were applied.

The least obvious part of the solution process involves formation of the global stiffness matrix by accumulating the contribution of each member. A truss having n nodes and m members will produce a system of $2n$ equations to be solved for a generalized displacement vector of the form

$$U = \begin{bmatrix} u_1 \\ v_1 \\ u_2 \\ v_2 \\ \vdots \\ u_n \\ v_n \end{bmatrix}$$

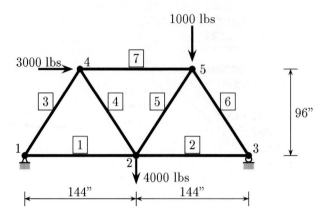

Figure 8.13. Example Truss Problem

Hence a member involving nodes i and j will influence rows $2i-1$, $2i$, $2j-1$, and $2j$. MATLAB has excellent facilities to deal with the type of problem discussed here because vectors can be used as subscripts in the assembly process, and intrinsic matrix commands are provided for solving the simultaneous equations.

8.4.1 Program to Analyze a Pin-Connected Truss

A computer program was written to compute member forces and nodal reactions in a general truss having known forces applied at selected nodes. The code, along with the output for the example problem shown in Figure 8.13, is contained in Section 8.4.3. A description of the code is summarized in Table 8.4. The allowable nodal displacement conditions handle nodes where both displacement components are known, as well as situations involving a roller which moves either in a horizontal direction or in a vertical direction.

8.4.2 Exercises

1. Write a function which plots both the undeflected and deflected geometry for the truss. For the deflected geometry utilize a scaling parameter to exaggerate the displacements.

2. In Section 8.4 the process of incorporating a roller on an inclined plane was discussed (see Figure 8.12). Make the necessary enhancements to program **trussrun** to incorporate the capability to define roller boundaries on inclined planes.

8.4. Analysis of Pin-Connected Trusses

Table 8.4. Description of Code in Example **trussrun**

Routine	Line	Operation
trussrun		script file to execute program.
	41-54	define the input parameters.
	57	assemble the global stiffness matrix.
	60-61	solve for the displacements and reactions.
	64	calculate the member forces.
	66-106	output the input parameters.
	107-121	output nodal displacements and reactions.
elmstf		function which computes the stiffness matrix for a single element.
	19	calculate the deformed length of the element.
	20	calculate some constants.
	21	create member stiffness matrix.
asmstf		function which assembles the individual member stiffness into a global stiffness matrix.
memfor		function which uses the member properties and the global displacement vector to compute member forces.
solve		function which modifies the global stiffness matrix to impose displacement constraints and solve for the global displacement vector.

8.4.3 Program Output and Code

Output from Example trussrun

```
Analysis of Pin-Connected Truss
-------------------------------

Member Information:
    #       E            A         i    j
   --- ------------ ------------ ---- ----
    1   3.0000e+07   1.00000e+01   1    2
```

```
     2    3.0000e+07   1.00000e+01      2    3
     3    3.0000e+07   1.00000e+01      1    4
     4    3.0000e+07   1.00000e+01      4    2
     5    3.0000e+07   1.00000e+01      2    5
     6    3.0000e+07   1.00000e+01      3    5
     7    3.0000e+07   1.00000e+01      4    5
```

Node list:
```
     #         x            y
    ---   ------------  ------------
     1    0.00000e+00   0.00000e+00
     2    1.44000e+02   0.00000e+00
     3    2.88000e+02   0.00000e+00
     4    7.20000e+01   9.60000e+01
     5    2.16000e+02   9.60000e+01
```

Boundary conditions:
```
     #    Condition     Value
    ---  ------------  ------------
     1       u         0.00000e+00
     1       v         0.00000e+00
     3       v         0.00000e+00
     4       Fx        3.00000e+03
     2       Fy       -4.00000e+03
     5       Fy       -1.00000e+03
```

Nodal Displacements:
```
     #         u            v
    ---  ------------  ------------
     1    4.36557e-18  -0.00000e+00
     2    1.89000e-03  -5.78250e-03
     3    3.24000e-03  -4.06999e-19
     4    3.75833e-03  -3.60000e-03
     5    1.41833e-03  -3.71000e-03
```

Nodal Reactions:
```
     #        Fx            Fy
    ---  ------------  ------------
     1   -3.00000e+03   1.25000e+03
     2   -1.81899e-12  -4.00000e+03
     3    0.00000e+00   3.75000e+03
     4    3.00000e+03  -1.81899e-12
     5   -2.72848e-12  -1.00000e+03
```

8.4. Analysis of Pin-Connected Trusses

Script File trussrun

```
 1: % Example: trussrun
 2: % ~~~~~~~~~~~~~~~~
 3: % This program performs static analysis of
 4: % plane truss structures using the stiffness
 5: % method. The truss elements are used to form
 6: % the global stiffness matrix of a structure.
 7: % Displacements, member forces and reaction
 8: % forces are obtained by solving the equations
 9: % which satisfy the prescribed boundary
10: % conditions and applied nodal forces.
11: % Data is defined in the declaration statements
12: % below, where:
13: %
14: % For each element:
15: %    Emod   - Modulus of elasticity
16: %    Area   - cross-sectional area
17: %    inode  - nodal number of i-end
18: %    jnode  - nodal number of j-end
19: % For each node:
20: %    x      - x coordinate
21: %    y      - y coordinate
22: % The specified D.O.F. and applied nodal loads:
23: %    idu    - nodal number of specified
24: %             x-displacement
25: %    u      - specified x-displacement
26: %    idv    - nodal number of specified
27: %             y-displacement
28: %    v      - specified y-displacement
29: %    idfx   - nodal number of applied load in
30: %             x-direction
31: %    fx     - applied load in x-direction
32: %    idfy   - nodal number of applied load in
33: %             y-direction
34: %    fy     - applied load in y-direction
35: %
36: % User m functions required:
37: %    asmstf, elmstf, memfor, solve
38: %-------------------------------------------------
39:
40: clear;
41: Problem=1;
```

```
42: if Problem == 1
43:    %...Simply supported 7-bar, 5-node plane truss
44:    Emod  =[3.e7 3.e7 3.e7 3.e7 3.e7 3.e7 3.e7];
45:    Area  =[  10   10   10   10   10   10   10];
46:    inode =[   1    2    1    4    2    3    4];
47:    jnode =[   2    3    4    2    5    5    5];
48:    x     =[   0  144  288   72  216];
49:    y     =[   0    0    0   96   96];
50:    idu   =[1];    idfx=[4];
51:    u     =[0];    fx  =[3000];
52:    idv   =[1 3];  idfy=[2 5];
53:    v     =[0 0];  fy  =[-4000 -1000];
54: end
55:
56: %...Assemble the global stiffness matrix
57: Stiff=asmstf(x,y,Area,Emod,inode,jnode);
58:
59: %...Displacement and reaction solutions
60: [Disp,Reactions]= ...
61:    solve(Stiff,idu,u,idv,v,idfx,fx,idfy,fy);
62:
63: %...Member force solutions
64: Mforces=memfor(Disp,x,y,Area,Emod,inode,jnode);
65:
66: fprintf( ...
67:    '\n\nAnalysis of Pin-Connected Truss');
68: fprintf( ...
69:    '\n-------------------------------');
70: fprintf('\n\nMember Information:');
71: fprintf('\n      #           E            A      ');
72: fprintf('    i    j');
73: fprintf('\n            ---  ------------ ------------');
74: fprintf('  ----  ----');
75: for i=1:length(Emod)
76:    fprintf('\n    %4.0f %12.4e %12.5e', ...
77:       i,Emod(i),Area(i));
78:    fprintf(' %4.0f %4.0f',inode(i),jnode(i));
79: end
80: fprintf('\n\nNode list:');
81: fprintf('\n      #         x             y');
82: fprintf('\n            ---  ------------ ------------');
83: for i=1:length(x)
84:    fprintf('\n    %4.0f %12.5e %12.5e', ...
85:       i,x(i),y(i));
```

8.4. Analysis of Pin-Connected Trusses

```
86:  end
87:
88:  fprintf('\n\nBoundary conditions:');
89:  fprintf('\n    #    Condition    Value');
90:  fprintf('\n   ---  -----------  ------------');
91:  for i=1:length(idu)
92:     fprintf('\n   %4.0f      u      %12.5e', ...
93:        idu(i),u(i));
94:  end
95:  for i=1:length(idv)
96:     fprintf('\n   %4.0f      v      %12.5e', ...
97:        idv(i),v(i));
98:  end
99:  for i=1:length(idfx)
100:    fprintf('\n   %4.0f      Fx     %12.5e', ...
101:       idfx(i),fx(i));
102: end
103: for i=1:length(idfy)
104:    fprintf('\n   %4.0f      Fy     %12.5e', ...
105:       idfy(i),fy(i));
106: end
107: fprintf('\n\nNodal Displacements:');
108: fprintf('\n    #        u              v');
109: fprintf('\n   ---  ------------  ------------');
110: for i=1:2:length(Disp)-1
111:    fprintf('\n   %4.0f %12.5e %12.5e', ...
112:       fix(i/2)+1,Disp(i),Disp(i+1));
113: end
114: fprintf('\n\nNodal Reactions:');
115: fprintf('\n    #       Fx            Fy');
116: fprintf('\n   ---  ------------  ------------');
117: for i=1:2:length(Reactions)-1
118:    fprintf('\n   %4.0f %12.5e %12.5e', ...
119:       fix(i/2)+1,Reactions(i),Reactions(i+1));
120: end
121: fprintf('\n');
```

Function elmstf

```
1: function k=elmstf(x,y,Area,Emod,inode,jnode)
2: %
3: % k=elmstf(x,y,Area,Emod,inode,jnode)
```

```
 4: %
 5: % This function forms the stiffness matrix
 6: % for a truss element.
 7: %
 8: % x,y         - global nodal coordinate vectors
 9: % Area        - member area vector
10: % Emod        - modulus of elasticity vector
11: % inode,jnode - indices of the member ends
12: %
13: % k           - member stiffness matrix
14: %
15: % User m functions called:   none
16: %-----------------------------------------
17:
18: i=inode; j=jnode;
19: xx=x(j)-x(i); yy=y(j)-y(i); L=norm([xx,yy]);
20: c=xx/L; s=yy/L; cc=c*c; cs=c*s; ss=s*s;
21: k=[[cc,cs];[cs,ss]]; k=Area*Emod/L*[k,-k;-k,k];
```

Function asmstf

```
 1: function stif=asmstf(x,y,Area,Emod,inode,jnode)
 2: %
 3: % stif=asmstf(x,y,Area,Emod,inode,jnode)
 4: %
 5: % This function assembles the global stiffness
 6: % matrix for a plane truss structure.  The
 7: % stiffness matrix has an order equal to twice
 8: % the number of nodes.
 9: %
10: % x,y         - global nodal coordinate vectors
11: % Area        - member area vector
12: % Emod        - modulus of elasticity vector
13: % inode,jnode - indices of the member ends
14: %
15: % stif - global stiffness matrix
16: %
17: % User m functions called:   elmstf
18: %-----------------------------------------
19:
20: numnod=length(x); numelm=length(Area);
21: i=inode(:); j=jnode(:);
```

8.4. Analysis of Pin-Connected Trusses

```
22: stif=zeros(2*numnod); ij=[2*i-1,2*i,2*j-1,2*j];
23: %...Place the stiffness matrix for individual
24: %...members at proper positions in the global
25: %...matrix
26: for k=1:numelm, kk=ij(k,:);
27:   stif(kk,kk)=stif(kk,kk)+ ...
28:     elmstf(x,y,Area(k),Emod(k),i(k),j(k));
29: end
```

Function memfor

```
 1: function ff=memfor(dd,x,y,Area,Emod,inode,jnode)
 2: %
 3: % ff=memfor(dd,x,y,Area,Emod,inode,jnode)
 4: % ~~~~~~~~~~~~~~~~~~~~~~~~~~~~~~~~~~~~~~
 5: % This function computes the axial force in
 6: % all members as a function of the member
 7: % properties and the global displacement vector.
 8: %
 9: % dd           - global displacement vector
10: % x,y          - global nodal coordinate vectors
11: % Area         - member area vector
12: % Emod         - modulus of elasticity vector
13: % inode,jnode  - indices of the member ends
14: %
15: % ff           - member force vector
16: %
17: % User m functions called:  none
18: %-----------------------------------------------
19:
20: x=x(:); y=y(:); a=Area(:); e=Emod(:);
21: i=inode(:); j=jnode(:);
22: xx=x(j)-x(i); yy=y(j)-y(i);
23: L=sqrt(xx.^2+yy.^2);
24: cs=xx./L; sn=yy./L; eaL=(a.*e)./L;
25: iu=dd(2*i-1); iv=dd(2*i);
26: ju=dd(2*j-1); jv=dd(2*j);
27: ff=eaL.*((ju-iu).*cs+(jv-iv).*sn);
```

Function solve

```
 1: function [disp,react]=solve ...
 2:    (stiff,idu,u,idv,v,idfx,fx,idfy,fy)
 3: %
 4: % [disp,react]=solve(stiff,idu,u,idv, ...
 5: %                   v,idfx,fx,idfy,fy)
 6: %~~~~~~~~~~~~~~~~~~~~~~~~~~~~~~~~~~~~~
 7: % This function solves for the nodal
 8: % displacements and reactions.
 9: %
10: % stiff    - the stiffness matrix
11: % idu,u    - nodal indices and magnitudes of all
12: %            known horizontal displacements
13: % idv,v    - nodal indices and magnitudes of all
14: %            known vertical displacements
15: % idfx,fx  - nodal indices and magnitudes of all
16: %            nonzero known horizontal nodal
17: %            forces
18: % idfy,fy  - nodal indices and magintudes of all
19: %            nonzero known vertical nodal forces
20: %
21: % disp     - computed nodal displacements
22: % react    - computed nodal forces
23: %
24: % User m functions called:  none
25: %----------------------------------------
26:
27: %...Form and solve the system:  STIFF*DISP=F
28: ns=size(stiff,1); f=zeros(ns,1);
29: ndc=size([u(:);v(:)]);
30: irow=[2*idu(:)-1;2*idv(:)];
31: %...Form vector f involving known forces and
32: %...displacements
33: f([2*idfx(:)-1;2*idfy(:)])=[fx(:);fy(:)];
34: f(irow)=[u(:);v(:)];
35: %...Modify rows of the stiffness matrix
36: %...corresponding to known displacements
37: diagv=zeros(ns,1); diagv(irow)=ones(ndc,1);
38: s=stiff; s(irow,:)=zeros(ndc,ns);
39: s=s+diag(diagv);
40: %...Solve for displacements and nodal forces
41: disp=s\f; react=stiff*disp;
```

Appendix A

Utility Routines

A.1 A System Dependent Plot Save Function

The collection of programs included in this book do not require any routines which are system dependent. However, many of the routines contain a "commented out" call to function **genprint**. This routine provides a convenient mechanism to generate plot files which are system/user dependent. The version of **genprint** included uses MATLAB's **print** command to generate a plot file in encapsulated postscript format. If the file name specified already exists then **genprint** erases the file prior to creating the plot file. The command to erase a file is system dependent and routine **genprint** contains the necessary logic for both MS-DOS systems and SGI UNIX systems.

A.1.1 Program Output and Code

Function genprint

```
 1: function genprint(fname,append)
 2: %
 3: % genprint(fname,append)
 4: % ~~~~~~~~~~~~~~~~~~~~~~
 5: % This function saves a plot to a file.  If
 6: % the file exists, it is erased first unless
 7: % the append option is specified.
 8: %
 9: % fname   - name of file to save plot to
10: %           without a filename extension
11: % append - optional, if included plot is
12: %           appended to file fname
13: %
14: % SYSTEM DEPENDENT ROUTINE
15: %
16: % User m functions called:   none
```

285

```
17: %------------------------------------------
18:
19: %...Define these appropriately
20: ext=['.eps'];  % filename extension to use
21: opt=['eps'];   % option for print command
22:
23: %...Append extension to filename
24: file_name=[fname,ext];
25:
26: %...Determine computer type
27: system_type=computer;
28:
29: %...Use correct command for different systems
30: if strcmp(system_type(1:2),'PC')
31:    erase_cmd=['delete ', file_name];
32: elseif strcmp(system_type(1:3),'SGI')
33:    erase_cmd=['!rm ', file_name];
34: else
35:    disp(' ');
36:    disp('Unknown system type in genprint');
37:    break;
38: end
39:
40: % Save to encapsulated postscript file
41: if nargin == 1
42:    if exist(file_name)==2
43:       eval(erase_cmd);
44:    end
45:    eval(['print -d',opt,' ',fname]);
46: else
47:    eval(['print -d',opt,' -append ',fname]);
48: end
```

A.2 Polygonal Representation of a Circle

Many of the methods presented in this book are quite powerful for studying problems which have complex geometries. This power was achieved by developing algorithms which can analyze cross sections defined by an n-sided polygon. Frequently, a cross section may be circular or contain boundary segments which are circular segments. Function **circle** generates an n-sided polygonal approximation for a circle. The circle is constructed using n equal angle increments.

A.2.1 Program Output and Code

Function circle

```
1: function [x,y]=circle(no_pts,x0,y0,radius)
2: %
3: % [x,y]=circle(no_pts,x0,y0,radius)
4: % ~~~~~~~~~~~~~~~~~~~~~~~~~~~~~~~~
5: % This function constructs the (x,y)
6: % coordinates necessary to draw a circle.
7: %
8: % no_pts - number of points to construct
9: %          circle with
10: % x0,y0 - center coordinate of circle
11: % radius - radius of circle
12: %
13: % Notes:
14: %   A typical use of this function is:
15: %         [x,y]=circle(50,2,3,6);
16: %         plot(x,y);
17: %         axis('equal');
18: %   The axis command causes the aspect ratio of
19: %   the circle to be correct.
20: %
21: % User m functions called:  none.
22: %-----------------------------------------------
23:
24: theta=linspace(0,2*pi,no_pts+1);
25: x=radius*cos(theta); x=x+x0;
26: y=radius*sin(theta); y=y+y0;
```

A.3 Function to Flip Angle Measures

Several of the problems presented in this book involve angle measures to planes of stress or strain. For this set of problems any angle in the second quadrant can be replaced with an angle in the fourth quadrant. The angle in the fourth quadrant is 180° removed from the angle in the second quadrant. A similar process can be used to transform an angle in the third quadrant to an angle in the first quadrant. Angles in the first and fourth quadrant have the range $-90°$ to $+90°$ when measured from the positive x-axis. The use of angle measurements which are only in the first and fourth quadrants is employed to simplify the presentation of results. Function **flpang** returns the equivalent first or fourth quadrant angle for a specified angle.

A.3.1 Program Output and Code
Function flpang

```
 1: function [NewAngle]=flpang(Angle,DegRad)
 2: %
 3: % [NewAngle]=flpang(Angle,DegRad)
 4: % ~~~~~~~~~~~~~~~~~~~~~~~~~~~~~~
 5: % This function transforms angle measures
 6: % to always be in the 1st and 4th quadrants.
 7: %
 8: % Angle    - angle to transform
 9: %            (positive measured CCW from
10: %            positive x-axis)
11: % DegRad   - ~=0, Angle/NewAngle are in degrees
12: %            =1, Angle/NewAngle are in radians
13: %
14: % NewAngle - transformed angle
15: %
16: % User m functions called:  none
17: %------------------------------------------------
18:
19: if DegRad == 1
20:    %...Angles in radians
21:    A1=pi/2; A2=pi; A3=3*pi/2; A4=2*pi;
22: else
23:    %...Angles in degrees
24:    A1=90; A2=180; A3=270; A4=360;
25: end
26: %...First, make between -360 and +360 degs
27: NewAngle=rem(Angle,A4);
28: %...Now, between +90 and -90
29: if abs(NewAngle)>A1 & abs(NewAngle)<A2
30:    NewAngle=rem(NewAngle,A1)-A1*sign(NewAngle);
31: elseif abs(NewAngle)==A2
32:    NewAngle=0;
33: elseif abs(NewAngle)>A2 & abs(NewAngle)<A3
34:    NewAngle=rem(NewAngle,A1);
35: elseif abs(NewAngle)==A3
36:    NewAngle=A1;
37: elseif abs(NewAngle)>A3 & abs(NewAngle)<A4
38:    NewAngle=rem(NewAngle,A1)-A1*sign(NewAngle);
39: end
```

A.4 Function for Polygon Clipping

In this section a program is presented which can be used to clip a polygon along any arbitrary plane. The techniques employed exploit several characteristics of the equation of a line. Therefore, a review of these relationships is provided. The relationships are then used to implement the program to clip a polygon along an arbitrary plane.

A.4.1 Equation of a Line

The most common method of representing the equation of a line is the **slope-intercept** form which is
$$y = mx + b$$
where m is the slope of the line, b is the y-intercept, and (x, y) is the coordinate of any point on the line. The routines used for polygon clipping require the line be represented in an alternate form, or
$$\alpha + \beta x + \gamma y = 0$$
where α, β, and γ are real numbers with β and γ not both zero. This is known as an **equation of the first degree** [25]. This equation has an analogous slope-intercept form which can be written as
$$y = \left(-\frac{\beta}{\gamma}\right)x + \left(-\frac{\alpha}{\gamma}\right)$$
If $\gamma = 0$ the equation reduces to an equation of a line parallel to the y-axis, or
$$x = \left(-\frac{\alpha}{\beta}\right)$$
Finally, the **normal form** of the equation of a line is written as
$$\frac{\alpha}{\pm\sqrt{\beta^2 + \gamma^2}} + \frac{\beta}{\pm\sqrt{\beta^2 + \gamma^2}}x + \frac{\gamma}{\pm\sqrt{\beta^2 + \gamma^2}}y = 0$$
with the sign of the radical chosen to be the same as the sign of γ. The normal form has some important properties which are shown in Figure A.1. The angle θ measures the angle between the x-axis and the normal vector $\hat{n}$. (The direction of the normal vector is always upward except when it coincides with the x-axis.) The coefficients β and γ of the normal form are related to θ by
$$\frac{\beta}{\pm\sqrt{\beta^2 + \gamma^2}} = \cos\theta \qquad \frac{\gamma}{\pm\sqrt{\beta^2 + \gamma^2}} = \sin\theta$$
and the perpendicular distance d between the line P and the origin is related to α by
$$\frac{\alpha}{\pm\sqrt{\beta^2 + \gamma^2}} = -d$$

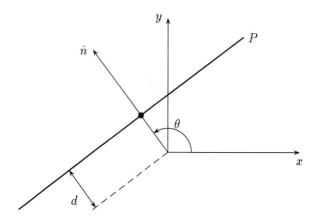

Figure A.1. Normal Form of a Line

and d is positive when the origin is below line P. Therefore, the normal form can also be expressed as
$$-d + (\cos\theta)x + (\sin\theta)y = 0$$

The clipping line P depicted in Figure A.1 divides the x-y plane into two distinct regions. Therefore, if the line is not parallel to the y-axis, every point is either above the line, below the line, or on the line[1]. Recall that the equation of a line can be written as
$$\alpha + \beta x + \gamma y = 0$$
Then for any point (x_a, y_a) when $\gamma > 0$[2] one of the following is true [25]:

$$\begin{aligned}
\alpha + \beta x_a + \gamma y_a &= 0 &&\text{point is on the line} \\
\alpha + \beta x_a + \gamma y_a &> 0 &&\text{point is above the line} \\
\alpha + \beta x_a + \gamma y_a &< 0 &&\text{point is below the line}
\end{aligned}$$

A.4.2 Program to Clip Arbitrary Polygon

The program to clip an arbitrary polygon requires that the clipping line be specified in normal form with the normal vector directed away from the part of the polygon

---

[1] For lines parallel to the y-axis the points are either on the left of the line, right of the line, or on the line.

[2] By reversing the sign of each term in the equation the meaning of above and below is also reversed.

A.4. Function for Polygon Clipping

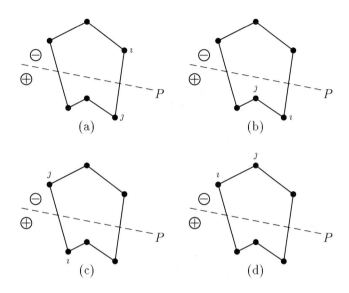

Figure A.2. Clipping Permutations

which is to be kept. The clipping process them becomes a matter of considering whether the clipping line intersects with a side of the polygon. This can be performed systematically by considering each side of the polygon in a step by step process and using the inequality relationships presented in the previous section. If the clipping line intersects the polygon[3] then one of four conditions is possible as shown in Figure A.2. Table A.1 summarizes each condition and the necessary action. The clipping process is straight forward and only requires careful attention to maintaining a continuously changing list of points which form the new "clipped" polygon.

---

[3] If the clipping line does not intersect the polygon then either all the points are eliminated or all the points kept.

Case	i	j	Action
(a)	inside $(-)$	outside $(+)$	keep point i, find point of intersection, and insert this point in new list of boundary nodes
(b)	outside $(+)$	outside $(+)$	discard point i and proceed to next boundary segment
(c)	outside $(+)$	inside $(-)$	discard point i, find point of intersection, and insert this point in new list of boundary nodes
(d)	inside $(-)$	inside $(-)$	keep point i and proceed to next boundary segment

Table A.1. Clipping Permutations

A.4.3 Program Output and Code

Function clip

```
 1: function [NoNewPts,IsetClipPoly,NoInterpPts, ...
 2:    Xclip,Yclip]=clip(Xclip,Yclip,NoClipPts, ...
 3:    Line,Epsilon)
 4: %
 5: % [NoNewPts,IsetClipPoly,NoInterpPts,Xclip, ...
 6: %    Yclip]=clip(Xclip,Yclip,NoClipPts, ...
 7: %    Line,Epsilon)
 8: % ~~~~~~~~~~~~~~~~~~~~~~~~~~~~~~~~~~~~~~~~
 9: % This function clips a reference polygon
10: % using a line defined in vector Line.
11: %
12: % Xclip,Yclip   - coordinates of reference
13: %                 polygon on input
14: % NoClipPts    - length of Xclip/Yclip
15: % Line         - vector containing coefficients
16: %                 of line equation in form:
17: %                 Line(1)+Line(2)*X+Line(3)*Y=0
18: %                 The normal vector to this
19: %                 line should be directed away
20: %                 from the region to be kept.
21: % Epsilon      - for determining zero values
22: %
23: % NoNewPts     - length of Xclip/Yclip on
24: %                 output
25: % IsetClipPoly - vector containing the
```

A.4. Function for Polygon Clipping

```
26: %                       indices for the clipped
27: %                       polygon in correct order
28: %  NoInterpPts    -  number of new points
29: %                       defined by interpolation
30: %  Xclip,Yclip    -  coordinates of clipped
31: %                       polygon on output
32: %
33: %  User m functions called:  inout, intrsect
34: %-------------------------------------------------
35:
36: %...Initialize
37: iPtConsider=1; iStartPoint=1; NoInterpPts=0;
38: NoNewPts=0; IdxNewPt=NoClipPts+1;
39: [IsFptIn]=inout(Xclip(iPtConsider), ...
40:                 Yclip(iPtConsider),Line);
41:
42: %...Loop on each side
43: for j=1:NoClipPts
44:   iStartPoint=j; iPtConsider=j+1;
45:   if j == NoClipPts, iPtConsider=1; end;
46:
47:   [IsNptIn]=inout(Xclip(iPtConsider), ...
48:                 Yclip(iPtConsider),Line);
49:
50:   if IsFptIn == 0 & IsNptIn == 1
51:     %...Current point is outside and next point
52:     %...is inside. Therefore, generate
53:     %...intermediate points.
54:     [IsectErr,Xclip,Yclip]=intrsect ...
55:       (Xclip,Yclip,Line,iStartPoint, ...
56:       iPtConsider,IdxNewPt,Epsilon);
57:     if IsectErr == 1
58:       error('\n\nError #1 in clip');
59:     end
60:     NoNewPts=NoNewPts+1;
61:     IsetClipPoly(NoNewPts)=IdxNewPt;
62:     IdxNewPt=IdxNewPt+1;
63:     NoInterpPts=NoInterpPts+1;
64:
65:   elseif IsFptIn == 1
66:     %...Current point inside, so save it
67:     NoNewPts=NoNewPts+1;
68:     IsetClipPoly(NoNewPts)=j;
69:     if IsNptIn == 0
```

```
70:        %...Next point is outside
71:        %...Generate intermediate point
72:        [IsectErr,Xclip,Yclip]=intrsect ...
73:          (Xclip,Yclip,Line,iStartPoint, ...
74:          iPtConsider,IdxNewPt,Epsilon);
75:        if IsectErr == 1
76:          error('\n\nError #2 in clip');
77:        end
78:        NoNewPts=NoNewPts+1;
79:        IsetClipPoly(NoNewPts)=IdxNewPt;
80:        IdxNewPt=IdxNewPt+1;
81:        NoInterpPts=NoInterpPts+1;
82:      end
83:    end
84:    IsFptIn=IsNptIn;
85: end
```

Function inout

```
 1: function [Flag]=inout(X,Y,Line)
 2: %
 3: % [Flag]=inout(X,Y,Line)
 4: % ~~~~~~~~~~~~~~~~~~~~~~
 5: % Determine if a point is above or below/on
 6: % a line.
 7: %
 8: % X,Y  - point to consider
 9: % Line - vector containing coefficients
10: %        of line equation in form
11: %        Line(1)+Line(2)*X+Line(3)*Y=0
12: %
13: % Flag - =0, point above line
14: %        =1, point below/on line
15: %
16: % NOTE: The theorem for determining whether
17: %       a point is above or below a line
18: %       assumes that Line(3)>0. By
19: %       reversing the signs of all the
20: %       coefficients of the line equation
21: %       the above/below sense is reversed.
22: %       (This is used advantageously in the
23: %       kern program.)
```

A.4. Function for Polygon Clipping

```
24: %
25: % User m functions called:  none
26: %-----------------------------------------
27:
28: Flag=0;
29: Product=Line(2)*X+Line(3)*Y+Line(1);
30: if Product <= 0, Flag=1; end;
```

Function intrsect

```
 1: function [IsectErr,Xclip,Yclip]=intrsect ...
 2:    (Xclip,Yclip,Line,iStartPoint, ...
 3:    iPtConsider,IdxNewPt,Epsilon)
 4: %
 5: % [IsectErr,Xclip,Yclip]=intrsect ...
 6: %    (Xclip,Yclip,Line,iStartPoint, ...
 7: %    iPtConsider,IdxNewPt,Epsilon)
 8: %
 9: % ~~~~~~~~~~~~~~~~~~~~~~~~~~~~~~~~~~~~~~~
10: % This function determines the point of
11: % intersection between a specified line
12: % and reference polygon using linear
13: % interpolation.
14: %
15: % Xclip,Yclip    - coordinates of reference
16: %                  polygon on input
17: % Line           - vector defining the clipping
18: %                  line in form:
19: %                  Line(1)+Line(2)*X+Line(3)*Y=0
20: % iStartPoint    - starting node in reference
21: %                  polygon segment
22: % iPtConsider    - ending node in reference
23: %                  polygon segment
24: % IdxNewPt       - index into Xclip/Yclip to
25: %                  store generated point
26: % Epsilon        - for determining zero values
27: %
28: % IsectErr       - error condition if set to one
29: % Xclip,Yclip    - updated coordinates of clipped
30: %                  polygon on output
31: %
32: % User m functions called:  none
```

```
33: %--------------------------------------------
34:
35: IsectErr=0;
36: C1=Line(2)*Xclip(iStartPoint)+ ...
37:    Line(3)*Yclip(iStartPoint);
38: C2=Line(2)*Xclip(iPtConsider)+ ...
39:    Line(3)*Yclip(iPtConsider)-C1;
40: C1=C1+Line(1);
41: if abs(C2) < Epsilon
42:   if abs(C1) < Epsilon
43:     %...Line is tangent to plane, return point
44:     %...halfway between
45:     xtmp=(Xclip(iStartPoint)+ ...
46:         Xclip(iPtConsider))/2;
47:     ytmp=(Yclip(iStartPoint)+ ...
48:         Yclip(iPtConsider))/2;
49:   else
50:     %...Line is parallel or in the plane
51:     IsectErr=1; return;
52:   end
53: else
54:   %...Ordinary situation
55:   C12=-C1/C2;
56:   xtmp=Xclip(iStartPoint)+ ...
57:       C12*(Xclip(iPtConsider)- ...
58:       Xclip(iStartPoint));
59:   ytmp=Yclip(iStartPoint)+ ...
60:       C12*(Yclip(iPtConsider)- ...
61:       Yclip(iStartPoint));
62: end
63: Xclip(IdxNewPt)=xtmp; Yclip(IdxNewPt)=ytmp;
```

Function refpoly

```
1: function [NoClipPts,Xclip,Yclip]=refpoly(x,y)
2: %
3: % [NoClipPts,Xclip,Yclip]=refpoly(x,y)
4: % ~~~~~~~~~~~~~~~~~~~~~~~~~~~~~~~~~~~~
5: % Define a large polygon to use as initial
6: % clipping polygon.
7: %
8: % x,y         - coordinates of user defined
```

A.4. Function for Polygon Clipping

```
 9: %                 geometry
10: %
11: % Xclip,Yclip - vectors containing the
12: %               coordinates of the clipping
13: %               polygon
14: % NoClipPts   - length of Xclip/Yclip
15: %
16: % User m functions called: none
17: %-----------------------------------------
18:
19: Xmin=min(x); Xmax=max(x);
20: Ymin=min(y); Ymax=max(y);
21:
22: %...Reference polygon to clip against
23: Xclip(1)=-10*abs(Xmin); Yclip(1)=-10*abs(Ymin);
24: Xclip(2)= 10*abs(Xmax); Yclip(2)=Yclip(1);
25: Xclip(3)=Xclip(2);      Yclip(3)= 10*abs(Ymax);
26: Xclip(4)=Xclip(1);      Yclip(4)=Yclip(3);
27: NoClipPts=4;
```

Appendix B

Description of MATLAB Commands

Table B.1. Partial List of MATLAB 4.xx Statements

\>\>	
Managing Commands and Functions	
help	Online documentation.
what	Directory listing of M-, MAT- and MEX-files.
type	List M-file.
lookfor	Keyword search through the help entries.
which	Locate functions and files.
demo	Run demos.
path	Control MATLAB's search path.
info	Information about MATLAB and The MathWorks.
Managing Variables and the Workspace	
who	List current variables.
whos	List current variables, long form.
load	Retrieve variables from disk.
save	Save workspace variables to disk.
clear	Clear variables and functions from memory.
pack	Consolidate workspace memory.
size	Size of matrix.
length	Length of vector.
disp	Display matrix or text.
Working with Files and the Operating System	
	continued on next page

cd	Change current working directory
dir	Directory listing.
delete	Delete file.
getenv	Get environment value.
!	Execute operating system command.
diary	Save text of MATLAB session.
Controlling the Command Window	
clc	Clear command window.
home	Send cursor home.
format	Set output format.
echo	Echo commands inside script files.
more	Control paged output in command window.
Starting and Quitting from MATLAB	
quit	Terminate MATLAB.
^C	local abort
startup	M-file executed when MATLAB is invoked.
matlabrc	Master startup M-file.
Operators and Special Characters	
+	Plus.
−	Minus.
*	Matrix multiplication.
.*	Array multiplication.
^	Matrix power.
.^	Array power.
kron	Kronecker tensor product.
\	Backslash or left division.
/	Slash or right division.
./	Array division.
:	Subscripting, vector generation.
()	Parentheses.
[]	Brackets.
.	Decimal point.
..	Parent directory.
...	Continuation.
,	Comma.
;	Semicolon.
%	Comment.
	continued on next page

!	Exclamation point.
'	Transpose and quote.
.'	Nonconjugated transpose.
=	Assignment.
==	Equality.
<>	Relational operators.
&	Logical AND.
|	Logical OR.
~	Logical NOT.
xor	Logical EXCLUSIVE OR.
Logical Functions	
exist	Check if variables or functions exist.
any	True if any element or vector is true.
all	True if all elements of vector are true.
find	Find indices of non-zero elements.
isempty	True for empty matrices.
isstr	True for text string.
Matrix Operators	
+	Addition.
−	Subtraction.
*	Multiplication.
/	Right division.
\	Left division.
^	Power.
'	Conjugate transpose.
Array Operators	
+	Addition.
−	Subtraction.
.*	Multiplication.
./	Right division.
.\	Left division.
.^	Power.
.'	Transpose.
Relational and Logical Operators	
<	less than.
<=	less than or equal.
>	greater than.

continued on next page

Description of MATLAB Commands

>=	greater than or equal.
==	equal.
~=	not equal.
&	AND.
|	OR.
~	NOT.
Special Characters	
[	Used to form vectors and matrices.
]	See [.
(	Arithmetic expression precedence.
)	See (.
,	Separate subscripts and function arguments.
;	End rows, suppress printing.
:	Subscripting, vector generation.
!	Execute operating system command.
MATLAB Programming Constructs	
function	Add new function.
eval	Execute string with MATLAB expression.
feval	Execute function specified by string.
global	Define global variable.
nargchk	Validate number of input arguments.
Control Flow	
if	Conditionally execute statements.
else	Used with **if**.
elseif	Used with **if**.
end	Terminate **if, for, while**.
for	Repeat statements a specific number of times.
while	Repeat statements an indefinite number of times.
break	Break out of **for** and **while** loops.
return	Return to invoking function.
error	Display message and abort function.
Interactive Input	
input	Prompt for user input.
keyboard	Invoke keyboard as if it were a script file.
menu	Generate menu of choices for user input.
pause	Wait for user response.
Elementary Matrices	
	continued on next page

zeros	Zeros matrix.
ones	Ones matrix.
eye	Identity matrix.
rand	Uniformly distributed random numbers.
randn	Normally distributed random numbers.
linspace	Linearly spaced vector.
logspace	Logarithmically spaced vector.
meshgrid	X and Y arrays for 3-D plots.
:	Regularly spaced vector.
Special Variables and Constants	
ans	Most recent answer.
eps	Floating point relative accuracy.
realmax	Largest floating point number.
realmin	Smallest positive floating point number.
pi	π, 3.1415926535897...
i,j	$\sqrt{-1}$, imaginary unit.
inf	∞, infinity.
NaN	Not-a-Number.
flops	Count of floating point operations.
nargin	Number of function input arguments.
nargout	Number of function output arguments.
computer	Computer type.
Time and Dates	
clock	Wall clock.
cputime	Elapsed CPU time.
date	Calendar.
etime	Elapsed time function.
tic, toc	Stopwatch timer function.
Matrix Manipulation	
diag	Create or extract diagonals.
fliplr	Flip matrix in the left/right direction.
flipud	Flip matrix in the up/down direction.
reshape	Change size.
rot90	Rotate matrix 90 degrees.
tril	Extract lower triangular part.
triu	Extract upper triangular part.
:	Index into matrix, rearrange matrix.

continued on next page

Description of MATLAB Commands

	Specialized Matrices
compan	Companion matrix.
hadamard	Hadamard matrix.
hankel	Hankel matrix.
hilb	Hilbert matrix.
invhilb	Inverse Hilbert matrix.
magic	Magic square.
pascal	Pascal matrix.
rosser	Classical symmetric eigenvalue test problem.
toeplitz	Toeplitz matrix.
vander	Vandermonde matrix.
wilkinson	Wilkinson's eigenvalue test matrix.
	Elementary Math Functions
abs	Absolute value.
acos	Inverse cosine.
acosh	Inverse hyperbolic cosine.
angle	Phase angle.
asin	Inverse sine.
asinh	Inverse hyperbolic sine.
atan	Inverse tangent.
atan2	Four quadrant inverse tangent.
atanh	Inverse hyperbolic tangent.
ceil	Round towards plus infinity.
conj	Complex conjugate.
cos	Cosine.
cosh	Hyperbolic cosine.
exp	Exponential.
fix	Round towards zero.
floor	Round towards minus infinity.
imag	Complex imaginary part.
log	Natural logarithm.
log10	Common logarithm.
real	Complex real part.
rem	Remainder after division.
round	Round towards nearest integer.
sign	Signum function.
sin	Sine.
	continued on next page

sinh	Hyperbolic sine.
sqrt	Square root.
tan	Tangent.
tanh	Hyperbolic tangent.
Specialized Math Functions	
bessel	Bessel function.
ellipj	Jacobian elliptic integral.
ellipke	Complete elliptic integral.
erf	Error function.
gamma	Gamma function.
rat	Rational approximation.
Matrix Analysis	
cond	Matrix condition number.
norm	Matrix or vector norm.
rcond	LINPACK reciprocal condition estimator.
rank	Number of linearly independent rows or columns.
det	Determinant.
trace	Sum of diagonal elements.
null	Null space.
orth	Orthogonalization.
rref	Reduced row echelon form.
Linear Equations	
/ and \	Linear equation solution.
chol	Cholesky factorization.
lu	Factors from Gaussian elimination.
inv	Matrix inverse.
qr	Orthogonal-triangular decomposition.
nnls	Non-negative least-squares.
pinv	Pseudoinverse.
Eigenvalues and Singular Values	
eig	Eigenvalues and eigenvectors.
poly	Characteristic polynomial.
hess	Hessenberg form.
qz	Generalized eigenvalues.
rdf2csf	Real block diagonal form to complex diagonal form.
cdf2rdf	Complex-diagonal form to real block diagonal form.

continued on next page

Description of MATLAB Commands

schur	Schur decomposition.
balance	Diagonal scaling to improve eigenvalue accuracy.
svd	Singular value decomposition.
Matrix Functions	
expm	Matrix exponential.
logm	Matrix logarithm.
sqrtm	Matrix square root.
funm	Evaluate general matrix function.
Basic Operations	
max	Largest component.
min	Smallest component.
mean	Average or mean value.
median	Median value.
std	Standard deviation.
sort	Sort in ascending order.
sum	Sum of elements.
prod	Product of elements.
cumsum	Cumulative sum of elements.
cumprod	Cumulative product of elements.
trapz	Numerical integration using trapezoidal method.
Finite Differences	
diff	Difference function and approximate derivative.
gradent	Approximate gradient.
del2	Five-point discrete Laplacian.
Correlation	
corrcoef	Correlation coefficients.
cov	Covariance matrix.
Fourier Transforms	
fft	Discrete Fourier transform.
fft2	Two-dimensional discrete Fourier transform.
ifft	Inverse discrete Fourier transform.
ifft2	Two-dimensional inverse discrete Fourier transform.
abs	Magnitude.
angle	Phase angle.
fftshift	Move zeroth lag to center of spectrum.
Polynomials	
	continued on next page

roots	Find polynomial roots.
poly	Construct polynomial with specified roots.
polyval	Evaluate polynomial.
polyvalm	Evaluate polynomial with matrix argument.
residue	Partial-fraction expansion (residues).
polyfit	Fit polynomial to data.
conv	Multiply polynomials.
deconv	Divide polynomials.
Data Interpolation	
interp1	1-D interpolation (1-D table lookup).
interp2	2-D interpolation (2-D table lookup).
interpft	1-D interpolation using FFT method.
griddata	Data gridding.
Function Functions - Nonlinear Numerical Methods	
ode23	Solve differential equations, low order method.
ode45	Solve differential equations, high order method.
quad	Numerically evaluate integral, low order method.
quad8	Numerically evaluate integral, high order method.
fmin	Minimize function of one variable.
fmins	Minimize function of several variables.
fzero	Find zero of function of one variable.
fplot	Plot function.
Elementary X-Y Graphs	
plot	Linear plot.
loglog	Loglog scale plot.
semilogx	Semi-log scale plot.
semilogy	Semi-log scale plot.
fill	Draw filled 2-D polygons.
Specialized X-Y Graphs	
polar	Polar coordinate plot.
bar	Bar graph.
stairs	Stairstep plot.
errorbar	Error bar plot.
hist	Histogram plot.
fplot	Plot function.
Graph Annotation	
title	Graph title.

continued on next page

Description of MATLAB Commands

xlabel	X-axis label.
ylabel	Y-axis label.
zlabel	Z-axis label for 3-D plots.
text	Text annotation.
gtext	Mouse placement of text.
grid	Grid lines.
Line and Area Fill Commands	
plot3	Plot lines and points in 3-D space.
fill3	Draw filled 3-D polygons in 3-D space.
Contour and Other 2-D Plots of 3-D Data	
contour	Contour plot.
contour3	3-D contour plot.
Surface and Mesh Plots	
mesh	3-D mesh surface.
meshc	Combination mesh/contour plot.
surf	3-D shaded surface.
surfc	Combination surf/contour plot.
waterfall	Waterfall plot.
Graph Appearance	
view	3-D graph viewpoint specification.
hidden	Mesh hidden line removal mode.
axis	Axis scaling and appearance.
3-D Objects	
cylinder	Generate cylinder.
sphere	Generate sphere.
Figure Window Creation and Control	
figure	Create figure (graph window).
gcf	Get handle to current figure.
clf	Clear current figure.
close	Close figure.
Axis Creation and Control	
subplot	Create axes in tiled positions.
axes	Create axes in arbitrary positions.
gca	Get handle to current axes.
cla	Clear current axes.
axis	Control axis scaling and appearance.
hold	Hold current graph.

continued on next page

Handle Graphics Objects	
figure	Create figure window.
axes	Create axes.
line	Create line.
text	Create text.
patch	Create patch.
surface	Create surface.
Handle Graphics Operations	
set	Set object properties.
get	Get object properties.
reset	Reset object properties.
delete	Delete object.
drawnow	Flush pending graphics events.
Hardcopy and Storage	
print	Print graph or save graph to file.
printopt	Configure local printer defaults.
orient	Set paper orientation.
General Character String Functions	
abs	Convert string to numeric values.
setstr	Convert numeric values to string.
isstr	True for string.
str2mat	Form text matrix from individual strings.
eval	Execute string with MATLAB expression.
String Comparison	
strcmp	Compare strings.
upper	Convert string to uppercase.
lower	Convert string to lowercase.
String to Number Conversion	
num2str	Convert number to string.
int2str	Convert integer to string.
str2num	Convert string to number.
sprintf	Convert number to string under format control.
sscanf	Convert string to number under format control.

Appendix C

List of MATLAB Routines with Descriptions

Table C.1. List of MATLAB Routines with Descriptions

Routine	Chapter	Description
arrows	5	This function draws stress arrows for stresses acting on a differential element.
asmstf	8	This function assembles the global stiffness matrix for a plane truss for program **trussrun**.
axdefl	8	This program determines the deflection in a cantilever beam subjected to a transverse load, concentrated moment, and axial load located at the free end.
cbeamex	6	This program calculates the stress distribution in the cross section of a curved beam having a symmetric polygonal cross section.
cbprop	6	This function calculates a new geometrical property, $\int (dx\, dy)/y$, required in program **cbeamex**.
circle	4, 6, A	This function generates a set of equally spaced points on the circumference of circle.
clip	6, A	This function clips a reference polygon using the specified line.
compbeam	8	This function determines flexural stresses and curvature in a beam having a rectangular cross section composed of several layers which are bonded together.
		continued on next page

Routine	Chapter	Description
constant	7	This function determines the constants of integration for the discontinuity functions used in program **singspan**.
elmstf	8	This function forms the stiffness matrix for a truss element for program **trussrun**.
findt	6	This function calculates the thickness of the section after a polygon has been clipped.
flbolts	2	This program determines the stresses in a flange coupling having multiple rings of bolts.
flpang	4, 5, A	This function transforms angle measurements in quadrants two and three to equivalent angle measurements in quadrants one and four.
genarea	6	This function returns the resulting polygon after a clipping operation on a polygon.
genprint	1-8, A	SYSTEM DEPENDENT function which saves a plot to an encapsulated postscript file.
hooke	2	This program determines either the stresses or strains in a plate using Hooke's law.
indaxial	3	This program calculates the internal forces and stresses in an axially loaded indeterminate member.
inertang	4	This function determines the inertia values at a specified rotation angle.
inout	6, A	This function determines if a point is above or below/on a line.
intrsect	6, A	This function determines the point of intersection between a specified line and reference polygon using linear interpolation.
japprox	4	This program calculates the polar moment of inertia for a circle approximated by an n-sided polygon.
kern	6	This program determines the kern for an axially loaded compression member described by a polygonal cross section.
memfor	8	This function computes the axial force in all members in a plane truss for program **trussrun**.

continued on next page

List of MATLAB Routines with Descriptions

Routine	Chapter	Description
multirod	3	This program analyzes the effects from a load applied to several rods suspending from a horizontal plane. The rods all connect at the point of application of the load.
mxnorex	6	This program determines the angles for the moment vector which produce an extreme value of stress in a polygonal cross section.
polarstr	5	This function creates a polar graph of the stresses acting on a differential element.
prinert	4	This function determines the principal inertia axes and values.
prop	4, 6	This function calculates the geometrical properties of a polygon.
propex	4	This program demonstrates the use of all the functions which can be employed to calculate the geometrical properties of polygonal cross sections.
prplot	5	This function plots a diffential element and the corresponding principal stresses.
prstress	5	This function calculates the principal stresses for a differential element.
prstrex	5	This program determines the principal stresses and corresponding angle on a differential element.
rbar1	3	This program determines the deflection of a rigid bar suspended by a set of cables and subjected to a vertical load. This version is vectorized.
rbar2	3	This program determines the deflection of a rigid bar suspended by a set of cables and subjected to a vertical load. This version is NOT vectorized.
refpoly	6, A	This function generates a large rectangle to use as a starting area for the determination of the kern. This area is clipped until the remaining area is the kern.
rosette	5	This program calculates the stresses resulting from measurements taked from three types of strain rosettes.
		continued on next page

Routine	Chapter	Description
setup	7	This function provides the input definitions for program **super**.
seval	7	This function evaluates discontinuity functions for a specified x position on the beam.
shcenter	6	This program calculates the shear center for an open thin-walled cross section.
shftprop	4, 6	This function uses the parallel-axis theorem to shift the geometrical properties to the centroidal axes.
shrcon	8	This program calculates the shear stresses in the bolts of an eccentric shear connection.
shrstr1	6	This program calculates the shear flow on a specified subarea of a cross section.
shrstr2	6	This program calculates the shear stress distribution in a polygonal cross section.
singspan	7	This program calculates the shear, moment, rotation, and deflection in a single span beam using discontinuity functions.
solve	8	This function solves for the nodal displacements and rections in a plane truss for program **trussrun**.
strangle	5	This function calculates the stresses for a specified angle of rotation for a differential element.
strangpl	5	This function plots a diffential element and the corresponding stresses on a rotated plane.
strglex	5	This program determines the stresses acting on specified plane of a differential element.
super	7	This program calculates the shear, moment, rotation, and deflection for a simply supported beam using an approximate technique based on superposition. Discontinuity functions are used for this analysis.
bimetal	8	This program determines the deflection in a cantilever beam constructed of of two materials which is subjected to a temperature variation. This program analyzes the affect of the coefficient of thermal expansion and modulus of elasticity.

continued on next page

List of MATLAB Routines with Descriptions

Routine	Chapter	Description
true	7	This function returns the true values of shear, moment, rotation, and deflection for comparison with the approximate values calculated by program **super**.
truss	1	Program which analyzes the effects from a load applied at the point of connection in a two rod truss.
trussrun	8	Program which performs static analysis of a plane truss using the stiffness method.
tworods	1	Program which analyzes the effects from a load applied at the point of connection of two rods suspended from a horizontal plane.

Bibliography

[1] I. O. Angell. *A Practical Introduction to Computer Graphics*. John Wiley & Sons, New York, 1981.

[2] Ferdinand P. Beer and E. Russell Johnston, Jr. *Mechanics of Materials*. McGraw Hill, 1981.

[3] A. P. Boresi and O. M. Sidebottom. *Advanced Mechanics of Materials*. John Wiley & Sons, New York, 1985.

[4] J. Cohen and T. Hickey. Two algorithms for determining volumes of convex polyhedra. *Journal of ACM*, 26(3), July 1979.

[5] Thomas F. Coleman and Charles Van Lon. *Handbook for Matrix Computations*. SIAM, 1988.

[6] W. H. Connolly. *Design of Prestressed Concrete Beams*. F. W. Dodge Corporation, New York, 1960.

[7] Nathan H. Cook. *Mechanics of Materials for Design*. McGraw-Hill, 1984.

[8] R. D. Cook and W. C. Young. *Advanced Mechanics of Materials*. Macmillan Publishing, New York, 1985.

[9] W. P. Creager, J. Justin, and J. Hinds. *Engineering for Dams, Volume II: Concrete Dams*. John Wiley & Sons, New York, 1945.

[10] M. Cyrus and J. Beck. Generalized two- and three-dimensional clipping. *Computers & Structures*, 3, 1978.

[11] J. P. Den Hartog. *Strength of Materials*. McGraw-Hill, 1949. Republished by Dover, 1961.

[12] A. C. Eberhardt and G. H. Williard. Calculating precise cross-sectional properties for complex geometries. *Computers in Mechanical Engineering*, September/October 1987.

[13] W. Flugge. *Handbook of Engineering Mechanics*. McGraw-Hill, 1962.

[14] Alfred M. Freudenthal. *Introduction to the Mechanics of Solids*. John Wiley & Sons, New York, 1966.

[15] C. E. Fuller and W. A. Johnston. *Applied Mechanics Volume II*. John Wiley & Sons, New York, 1919.

[16] James M. Gere and Stephen P. Timoshenko. *Mechanics of Materials*. Wadsworth, Inc., second edition, 1984.

[17] Y. Guyon. *Limit-State Design of Prestressed Concrete, Volume 1*. John Wiley & Sons, New York, 1972.

[18] Charles O. Harris. *Introduction to Stress Analysis*. MacMillan Company, New York, 1959.

[19] Archie Higdon, Edward H. Ohlsen, William B. Stiles, and John A. Weese. *Mechanics of Materials*. John Wiley & Sons, New York, second edition, 1967.

[20] Y. T. Lee and A. A. G. Requicha. Algorithms for computing the volume and other integral properties of solids, I. known methods and open issues. *Communications of the ACM*, 25(9), 1982.

[21] Y. Liang and B. A. Barsky. An analysis and algorithm for polygon clipping. *Communications of the ACM*, 26(11), 1983.

[22] J. A. Liggett. Exact formulae for areas, volumes and moments of polygons and polyhedra. *Communications in Applied Numerical Methods*, 4, 1988.

[23] J. Marin. Computing columns, footings and gates through moments of area. *Computers & Structures*, 18(2), 1984.

[24] J. C. McCormac. *Design of Reinforced Concrete*. Harper & Row Publishers, New York, 1978.

[25] Neal H. McCoy and Richard E. Johnson. *Analytic Geometry*. Holt, Rinehart and Winston, New York, 1955.

[26] W. McGuire. *Steel Structures*. Prentice-Hall, Inc., New Jersey, 1968.

[27] R. G. Miles and J. G. Tough. A method for the computation of inertial properties for general areas. *Computer-Aided Design*, 15(4), July 1983.

[28] A. Morley. *Theory of Structures*. Longmans, Green and Co., London, 1948.

[29] B. B. Muvdi and J. W. McNabb. *Engineering Mechanics of Materials*. Macmillan, 1984.

[30] William A. Nash. *Strength of Materials*. Schaum's Outline Series, McGraw-Hill, 3rd edition, 1967.

[31] Hermann K. P. Neubert. *Strain Gages: Kinds and Uses*. Macmillan, 1967.

[32] A. H. Nilson. *Design of Prestressed Concrete*. John Wiley & Sons, New York, 1978.

[33] J. T. Oden. *Mechanics of Elastic Structures*. McGraw-Hill, New York, NY, 1967.

[34] Gerner A. Olsen. *Elements of Mechanics of Materials*. Prentice-Hall, Inc., Englewood Cliffs, NJ, 1958.

[35] Mario Paz, C. Patrick Strehl, and Preston Scharder. Computer detemination of the shear center of open and closed sections. *Computers & Structures*, 6, 1976.

[36] C. C. Perry and H. R. Lissner. *The Strain Gage Primer*. McGraw-Hill, Inc., second edition, 1962.

[37] Egor Paul Popov. *Mechanics of Materials*. Prentice-Hall, Inc., second edition, 1976.

[38] R. J. Roark and W. C. Young. *Formulas for Stress and Strain*. McGraw-Hill, 1975.

[39] Samuel M. Selby, editor. *Standard Mathematical Tables*. The Chemical Rubber Company, Cleveland, Ohio, 1968.

[40] F. R. Shanley, editor. *Mechanics of Materials*. McGraw-Hill, 1967.

[41] Kermit Sigmon. *MATLAB Primer*. CRC Press, Inc., Boca Raton, Florida, 1994.

[42] Vaclav Skala. An efficient algorithm for line clipping by convex polygon. *Computer & Graphics*, 17(4), 1993.

[43] E. Sutherland and W. Hodgman. Reentrant polygon clipping. *Communications of the ACM*, 17(1), 1974.

[44] The MathWorks Inc. *MATLAB User's Guide*. The MathWorks, Inc., South Natick, Mass., 1994.

[45] The MathWorks Inc. *The Student Edition of MATLAB For MSDOS Personal Computers*. The MATLAB Curriculum Series. Prentice-Hall, Englewood Cliffs, NJ, 1994.

[46] S. Timoshenko. *Strength of Materials - Part I: Elementary Theory and Problems*. Robert E. Krieger Publishing, Huntington, NY, third edition, 1958.

[47] S. Timoshenko. *Strength of Materials - Part II: Advanced Theory and Problems.* Robert E. Krieger Publishing, Huntington, NY, third edition, 1958.

[48] S. Timoshenko and Gleason H. MacCullough. *Elements of Strength of Materials.* D. Van Nostrand Company, New York, second edition, 1940.

[49] Stephen P. Timoshenko. *History of Strength of Materials.* McGraw-Hill, 1953. Republished by Dover, 1983.

[50] H. B. Wilson and K. Deb. Inertial properties of tapered cylinders and partial volumes of revolution. *Computer Aided Design*, 21(7), September 1989.

[51] H. B. Wilson and D. S. Farrior. Computation of geometrical and inertial properties for general areas and volumes of revolution. *Computer Aided Design*, 8(8), 1976.

[52] H. B. Wilson and J. L. Hill. Volume properties and surface load effects on three dimensional bodies. Technical Report BER Report No. 266-241, Department of Engineering Mechanics, University of Alabama, Tuscaloosa, Alabama, 1980. U.S. Army Engineer Waterways Experiment Station, Vicksburg, MS, 1980.

[53] H. B. Wilson and L. H. Turcotte. Determining the kern for a compression member of general cross-section. *Advances in Engineering Software*, 17(2), 1993.

[54] H. B. Wilson and L. H. Turcotte. *Advanced Mathematics and Mechanics Applications Using MATLAB.* CRC Press, Inc., Boca Raton, Florida, 1994.

Index

area of cross section, 65
axial forces, 1

beam
    approximate analysis, 201
    bending stress, 173
    buckling, 254
    combined loadings, 254
    constructed of several materials, 261
    constructed of two materials, 263
    curved, 173
    deflection diagram, 199
    load diagram, 199
    moment diagram, 199
    nonhomogeneous, 261
    rotation diagram, 199
    shear diagram, 199
beam columns, 254
bolts
    shearing stress, 31, 246
buckling, 254

centroid, 65
circle
    generation of, 286
core, 143

dilatation, 24, 25
discontinuity functions, 199
    Macaulay, 199
    singularity, 199

first moment of area, 65, 175

flange coupling, 31
flexibility coefficient, 40
flexure formula, 133, 144
    generalized, 133, 144

geometrical properties, 65
    of polygons, 65

Heron's formula, 14
Hooke's Law, 24

indeterminate, 39

kern, 143, 144
kernel, 143
kernel points, 150

law of cosines, 1
limit zone, 143
line
    equation of, 289
    equation of first degree, 289
    normal form, 148, 289
    slope-intercept form, 289

Macaulay functions, 199
MATLAB Commands
    summary of, 298
MATLAB Routines
    arrows, 98, 114
    asmstf, 277, 282
    axdefl, 257, 259
    bimetal, 264, 267
    cbeamex, 178, 180

cbprop, 178, 183
circle, 79, 150, 287
clip, 150, 161, 292
compbeam, 264, 270
constant, 232, 244
elmstf, 277, 281
findt, 161, 172
flbolts, 34, 36
flpang, 98, 288
genarea, 161, 171
genprint, 285
hooke, 26, 27
indaxial, 60, 61
inertang, 76, 90
inout, 150, 161, 294
intrsect, 150, 161, 295
japprox, 78, 79, 85
kern, 150, 153
memfor, 277, 283
multirod, 42, 43
mxnorex, 136, 140
polarstr, 98, 117
prinert, 76, 89
propex, 74, 75, 81
prop, 76, 79, 87, 136, 150, 161
prplot, 98, 110
prstress, 97, 106
prstrex, 96, 102
rbar1, 48, 52
rbar2, 54
refpoly, 150, 296
rosette, 128, 130
setup, 212, 222
seval, 231, 243
shcenter, 188, 192
shftprop, 76, 88, 136, 150, 161
shrcon, 250, 251
shrstr1, 157, 160, 165
shrstr2, 158, 160, 167
singspan, 231, 237
solve, 277, 284
strangle, 97, 105
strangpl, 97, 108
strglex, 96, 101

super, 211, 217
true, 212, 224
trussrun, 277, 279
truss, 17, 20
tworods, 4, 9
modulus of elasticity, 24
modulus of rigidity, 24
moment of inertia, 65
    parallel-axis theorem, 66
    principal axes, 69
    rotation of, 68

parallel-axis theorem, 66, 185
plate analysis, 24
Poisson's ratio, 24
polar moment of inertia, 65, 66
polygon
    area, 70
    centroid, 70
    clipping, 289
    first moment, 70
    for curved boundaries, 78
    geometrical properties, 65
    integral of $dx\,dy/y$, 175
    moment of inertia, 70
    second moment, 70
principal stress, 93
product of inertia, 66

radius of gyration, 65, 66
reciprocity, 146
rosette, 92, 121, 122
    delta, 121, 124
    rectangular, 121, 123
    T-delta, 121, 126

S-polygon, 143
second moment of area, 65
shear center, 184
shear connection, 246
shear flow, 157
shear formula, 156
shear stress
    due to bending, 156

singularity functions, 199
stiffness coefficient, 46
stiffness matrix, 274
strain
    principal axes, 119
    rotation of, 92, 119
    transformation of, 92, 119
strain rosette, 92, 121
stress
    principal axes, 94
    rotation of, 92, 93
    transformation of, 92, 93
superposition, 199, 201

thermostat, 263
transfer formula, 66
transformation
    strain, 92, 119
    stress, 92
truss, 272
    pin-connected, 272

unit load method, 57

volumetric strain, 25

Three easy ways to receive more information on MATLAB:

- Fax this form to (508) 647-7101
- Mail this form to The MathWorks, Inc., P.O. Box 5936, Holliston, MA 01746-9752
- Send e-mail to *info@mathworks.com* and request kit KP108

Send me a *free* copy of the MATLAB® Product Catalog.

This catalog provides information on MATLAB, Toolboxes, SIMULINK®, Blocksets, and more.

I am currently a MATLAB user: ☐ Yes ☐ No

Computer Platform: ☐ PC or Macintosh ☐ UNIX workstation

For the fastest response, fax to (508) 647-7101 or send e-mail to *info@mathworks.com* and request KP108.

NAME
E-MAIL
TITLE
COMPANY/UNIVERSITY
DEPT. OR M/S
ADDRESS
CITY/STATE/COUNTRY/ZIP
PHONE
FAX
GRADUATION DATE IF APPLICABLE

R-BK-TUR/411v0/KP108